INTERNATIONAL ARCHITECTURE WATER LANDSCAPE

国际建筑滨水景观

高迪国际 HI-DESIGN PUBLISHING 编

中国林业出版社

图书在版编目（ＣＩＰ）数据

国际建筑滨水景观：汉英对照 / 高迪国际编；王丹等译． —— 北京：中国林业出版社，2013.9
　　ISBN 978-7-5038-7179-5

Ⅰ．①国… Ⅱ．①高… ②王… Ⅲ．①理水（园林）－景观设计－世界－图集 Ⅳ．① TU986.4-64

中国版本图书馆 CIP 数据核字（2013）第 209676 号

中国林业出版社 • 建筑与家居出版中心
责任编辑：纪亮 李丝丝
出版咨询：（010）8322 5283

出版：中国林业出版社 （100009 北京西城区德内大街刘海胡同 7 号）
网址：http://lycb.forestry.gov.cn/
电话：（010）8322 5283
发行：中国林业出版社
印刷：北京利丰雅高长城印刷有限公司
版次：2013 年 9 月第 1 版
印次：2013 年 9 月第 1 次
开本：8 开
印张：44
字数：400 千字
定价：780.00 元（USD140.00）

INTERNATIONAL ARCHITECTURE WATER LANDSCAPE

PREFACE I 序言一

WATERFRONTS
滨水区

Oceans, lakes and rivers have always been the focus for human activity. The water's edge has always been a special place for people, a place to gather and to settle, a place for work and for contemplation, a place of nourishment and a place of departure and arrival.

In nature, the waterfront marks the boundary between two worlds, the land and the water, and brings with it richness and diversity that is hard to compare.

The place where these great elements meet should always be treated with respect and dignity. Water connects and separates us, it beckons us to another world and challenges us to explore. The water's edge allows us to ponder another place where our imaginations can roam. It's a place of spirituality.

In modern times, these statements are both true and sadly neglected. Many water bodies have been exploited, depleted and dominated by one use or another. Polluted and devoid of life, water can also be an irritant, a reminder of our failures and our greed. The spirituality of the waterfront can only be fully attained when it has been repaired, when the water again breathes life.

All over the world, waterfronts are being reimagined, restored, re purposed and again given the respect and dignity they deserve. Through this process, water quality has dramatically improved along with aquatic life. Modern urban dwellers still understand the intrinsic uniqueness of being by the water. Waterfronts still pull at our hearts and take us to another place. Today waterfronts provide a vital spiritual resource in urban development.

Modern urban waterfronts should be seen as a place for communal gathering, a place of good health and of celebration. The modern waterfront can continue to feed us, literally and spiritually. Waterfront developments provide us with the opportunity to improve the water environment. To halt the process of degradation, to create the opportunity for nature at the water's edge and to provide the opportunity for man to witness, understand and experience this transformation.

In such contexts communities thrive. People come to live and enjoy the view. People gather and explore. They come for recreation, health and leisure.

Waterfronts are often rich in human memory, of past times, of communities and of sacred stories. Successful waterfront development draws on the genius loci of the place and provides a reflection of the past resulting in richness and authenticity often lacking in modern development.

Through careful and diligent planning waterfront communities can celebrate and remember the sacred value of the water's edge and all that it has provided for generations past.

　　海洋、湖泊和河流，自古以来都是人类活动的焦点场所。

　　滨水地带对人类有着特殊的意义：不但是绝佳的安居之所，也是众多产业、工业的发源之地；它更是滋养文化的温床，是探索与远航的起点，是守望与回归的故乡。人类在这里栖居、劳作、沉思、梦想。

　　大自然中，滨水区是陆地和水域两个世界的分界线，其自身的丰富性和多样性是无可比拟的。

　　作为这些伟大元素的交汇之处，滨水区应始终受到人类的尊重。水将我们联系在一起，也将我们分开；将我们引领到另一个世界，向我们发出探索的挑战。身临水畔，我们得以在此岸与彼岸间沉思、遐想，信马由缰。这里是一个灵性世界，浪漫之乡。

　　然而，在现代社会中，这些观点虽真实可见，却往往被人们忽视。许多水体被定为单一用途进行开发、利用并已经枯竭。水体被污染，水中生物资源殆尽。水，刺激着我们的欲望，提醒着我们因贪婪而导致的种种失败。只有重新修复水体，唤起水中的生命气息，滨水区才能真正回归灵性、回归美好。

　　在世界各地，滨水区正被重新规划、修复和定义，找回应得的尊重和尊严。在这一过程中，水质和水生生物得到了极大改善。现代的都市居民依然懂得傍水而居的独特价值。滨水区依然牵动着每个人的心灵，将我们带到另一个世界。在当代城市的开发过程中，滨水区为我们提供了一方独特的精神家园。

　　现代城市滨水区，应被视为公共聚集地，拥有健康的环境和可以进行庆祝活动的场所，无论在物质层面还是精神层面都能够继续滋养我们。滨水区的开发为我们提供了改善水环境的机会：阻止水体的退化，为水生态创造恢复生机的机会，也为人类创造见证、了解和经历这一转变的机会。

　　这样的水畔才是社区生活蓬勃发展的最佳环境。这样的水畔，才能吸引人们前来定居、游览、娱乐、探险。

　　人类对水畔的记忆往往是丰富而神圣的：包含着历史的过往，承载着宗族的传说。成功的滨水区开发项目，着眼于滨水区所在地的本土特色，致力于在设计中融汇当地的历史背景和文化，给予现代开发项目中通常所缺少的丰富性和独创性。

　　只有通过精心的规划，滨水社区才能植根于水畔的神圣价值所在，铭记它对一代又一代人的滋养，延承它博厚的灵性魅力。

Chris Sterry, Principal of PWL Partnership Landscape Architects Inc.
克里斯·斯特里（PWL 规划和景观设计有限公司负责人）

PREFACE II 序言二

WATER
水

More than half of our bodies are made up of water. Is this why people are so drawn to it for peace and relaxation?

Water is one of the most powerful landscape attractants in both public and private places, in cities and natural environments. It's visual and audible cues are immediately relaxing and often awe-inspiring.

Water has been and remains a key component within many of the most well-known and highly recognized landscapes throughout the world, both natural and manmade. Its appeal transcends all countries, religions, and cultural values. Man's relationship with water is inextricably linked, be it as a source of sustenance, a commercial conduit or simply a place of beauty and recreation.

It is true to say that we never tire of looking at, listening to, or being beside water. Varying immensely in the type of interface, whether man-made or natural, waterfront landscapes are always highly valued destinations.

As Landscape Architects and Master Planners, waterfront landscapes present us with some of the most interesting and challenging design exercises. Whether beachfront, lakefront, canal or natural waterway, each of these landscapes carries not only exciting opportunities but also specific sets of guidelines and considerations to maximize the end result for the final users of these special places.

Landscapes adjacent or associated with water enjoy one of the most beautiful settings, and water, whether active or passive, attracts us to play or rest, and with luck to live beside. For this reason landscapes adjacent to water enjoy a very high level of desirability, not just recreationally and culturally, but in a commercial sense as well. It is very true that water adds tremendous value to land whether it is directly adjacent, or if there is only a glimpse of water from afar.

Therefore designing waterfront landscapes and master planning landscapes around water, means having to consider many different goals. Some common goals include maximizing the length of the water-landscape interface, considering accessibility and visibility of the water body, creating safe entry and egress from the water where applicable, optimizing the opportunities to interact, enjoy and appreciate the waterfront landscape, managing the co-existence of public and private interfaces to water, and by doing these things elevating the value that waterfront landscapes give to a development or city.

PLACE Design Group has been involved in many different types of waterfront landscapes, with a number of our waterfront projects attracting international design recognition. We are always excited to master plan and design the many types of landscapes that involve water frontage, from hard-edged boardwalks and parkland promenades to softer natural interfaces. The juxtaposition of landscape and water requires us as designers to appreciate and support the sensitive environmental processes that occur between them, and to design waterfront landscapes that ensure both ecological and economic sustainability.

我们的身体有一半以上由水组成。这或许可以解释为什么我们会被它的宁谧与舒缓所深深吸引吧？

无论在公共场所还是私人空间，在城市还是自然环境中，水景都是最令人向往的景观之一。水的视觉和听觉效果，亦可营造轻松的氛围，亦可带来庄严肃穆的效果。

在世界各地，无论是自然景观还是人造景观，水在最具知名度和认可度的景观中一直是一个关键的组成部分。水的魅力超越了所有国家、宗教和文化的界限。无论作为食物的来源、商业的渠道，或只是一个供人们娱乐的优美场所，水向来与人类密不可分。

可以说，对于水，我们从未厌倦去欣赏它、聆听它或是居住在它身边。水滨的种类繁多，无论是人造的还是自然形成的，滨水景观一直以来都备受重视。

作为景观建筑师和城市规划师，滨水景观为我们呈现了一系列最有趣和最具挑战性的设计实践。无论是海滨、湖畔、运河还是自然水道，这每一处景观都为我们带来了宝贵的创作机会，同时也促使我们深入对指导方针和细节的思考，以最大程度地优化设计，使最终使用者受益。

景观，无论是与水相邻还是与水有关，都极富魅力。而水，无论是活跃的还是宁静的，都吸引着我们前去观赏、游玩、甚至有幸定居。因此，滨水景观的强大吸引力，不仅具有娱乐和文化价值，也有巨大的商业价值。无论陆地是临水、近水，还是只有一眼远水，都无疑会因水增值。

因此，对滨水景观的设计与总体规划，必须全方面考虑多种不同的目标。一些常见的规划目标包括：最大程度地扩大滨水景观的临界面长度，考虑水体的可及性和可视性，在适当位置构建水体景观的安全入口和出口，最大程度地优化人与滨水景观的互动，使人尽情欣赏、享用滨水景观，同时妥善管理公共和个人空间与水体的共存关系，提升滨水景观给开发项目和所在城市带来的长远价值。

普利斯设计咨询公司参加过多个不同类型的滨水景观设计项目，其中一些项目还赢得了国际设计界的认可。我公司致力于筹划和设计与滨水相关的多种类型景观，包括硬边缘木板路、公园漫步道以及更为柔和的自然界面。景观与水体的交叉要求我们的景观设计者充分考量二者之间的敏感环境进程，设计出具有生态和经济可持续性的滨水景观。

Lindsay Thorpe, Founder of PLACE DESIGN GROUP
琳赛·索普（普利斯设计咨询公司创始人）

CONTENTS 目录

BONDI TO BRONTE COAST WALK EXTENSION 008 ／邦代至勃朗特滨海路扩建工程	**HORNSBERGS STRANDPARK** 094 ／霍恩博格海滨公园
DARWIN WATERFRONT PUBLIC DOMAIN 016 ／达尔文海滨公共区域	**SENTOSA BOARDWALK** 102 ／圣淘沙跨海步行道
JACK EVANS BOAT HARBOUR - TWEED HEADS - STAGE 1 024 ／杰克·埃文斯船港——堤维德岬一期工程	**LYON RIVER BANK** 112 ／里昂河岸
KEELUNG MARITIME PLAZA 036 ／基隆海洋广场	**SCHINKELEILANDEN** 124 ／辛克尔小岛
CLOCK TOWER BEACH 052 ／钟楼海滩	**VICTORIA HARBOUR, DOCKLANDS** 134 ／达克兰的维多利亚港
TEL AVIV PORT PUBLIC SPACE REGENERATION 060 ／特拉维夫港口公共空间的改造	**SHENZHEN BAY COASTLINE PARK** 144 ／深圳湾海岸公园
TRAFFORD WHARF PROMENADE 066 ／特拉福德滨河长廊	**THE INTERVENTION IN THE EBRO BANKS** 158 ／埃布罗河畔的干预项目
BULCOCK BEACH FORESHORE REDEVELOPMENT 074 ／布卡克海滩重建项目	**SEEPLATZ WETTER** 172 ／威特市斯博拉茨
SCARBOROUGH BEACH URBAN DESIGN MASTER PLAN STAGES 1 & 4 082 ／斯卡伯勒海滩城市设计规划1、4阶段	**SOUTHEAST FALSE CREEK** 180 ／东南福溪可持续社区

JON STORM PARK 194 乔恩斯托姆公园	**YALIKAVAK PALMARINA** 278 雅勒卡瓦克海滨码头
STASSFURT 200 斯塔斯福特	**MAASBOULEVARD VENLO—A NEW URBAN AREA AT RIVER MEUSE IN VENLO** 288 芬洛马斯河大街——芬洛马斯河岸新城区
PORT COOGEE REDEVELOPMENT 208 库吉港再开发项目	**WILKES-BARRE LEVEE RAISING - RIVER COMMONS** 294 威尔克斯巴里堤防加高河流公地
DIAMOND TEAGUE PARK 216 钻石提格公园	**PRINCE ARTHURS LANDING, HUNDER BAY WATERFRONT** 300 桑德贝滨水景观，亚瑟王子码头
KANGAROO BAY 224 袋鼠湾	**GARDENS BY THE BAY** 314 滨海湾花园群
NANHAI CITIZEN'S PLAZA AND THOUSAND LANTERN PARK 234 南海市民广场和千灯湖公园	**NEWPORT QUAYS** 326 新港码头
OLYMPIC SCULPTURE PARK 246 奥林匹克雕塑公园	**PROMENADE SAMUEL-DE CHAMPLAIN** 334 萨缪尔·德·尚普兰滨水长廊
MOOLOOLABA FORESHORE STAGE 2A 256 穆卢拉巴前滩重建2A期工程	**PROMENADE DES ANGLAIS** 340 盎格鲁滨海路
NORTH WHARF PROMENADE, JELLICOE STREET AND SILO PARK 266 新西兰杰利科北部码头漫步长廊，杰利科大道和筒仓公园	**APPENDIX** 348 附录

LANDSCAPE ARCHITECT ASPECT Studios   CLIENT Waverley Council

BONDI TO BRONTE COAST WALK EXTENSION
/ NSW, AUSTRALIA

邦代至勃朗特滨海路扩建工程

AREA 515 m²   **PHOTOGRAPHER** Florian Groehn

The Bondi to Bronte coast walk is a part of the nationally significant 9km coastal walk from Sydney's South Head to Maroubra. The project resolves complex geotechnical, structural and heritage conditions to retain the significant cliff top heath community and the remarkable hanging swamps along the exposed sandstone platforms. A set of lookouts strung together by a light thread of walkway along the cliff tops of Sydney's east captures the sublime of the headlands, the sandstone outcrops and the vastness at the continent's edge.

Sensitivity to site is embodied in the simple materials and understated lightness of the walkway's design as it shifts and slides along the movement path to reveal the story of the cliff top landscape. Each of the five lookout points takes on its own distinctive form as the walkway cranks and fractures in response to the crystalline geology of the site and angular structure of Hawkesbury sandstone. Over remnant vegetation, the walkway's timber switches to a gridded fiberglass mesh to allow light and water to penetrate. Materials are selected for design quality, durability and sustainability. A limited use of balustrades balances risk with experience.

Site conditions were complex, and rigorous geotechnical planning was required to work with the rock outcrops and dyke intrusions of the site. In many areas piling was required to breech unstable fill slopes, and environmental sensitivities meant construction access was limited to a four meter corridor. The project originally arose from the need to preserve the historic Waverley Cemetery, one of Sydney's most culturally sensitive and spectacular cemeteries, from ongoing damage by the walk's annual 700,000 visitors.

邦代至勃朗特的滨海栈道属于悉尼南端至马诺布拉滨海路的一部分。这条全长9千米的栈道是澳大利亚举国闻名的滨海栈道。该项目解决了复杂的地质技术问题：既要克服现有地质结构和遗产状况的困难，又要保护重要的崖顶石楠植物群以及沿着裸露的砂岩分布的大面积陡坡沼泽。沿悉尼东部悬崖顶端修建的步道将一系列观景台串连起来，在那里可以欣赏到雄伟的陆岬、突出地面的砂岩以及大陆尽头浩瀚的海洋。

设计师对景点环境的体察细致入微。木栈道的设计用材简单，颜色清新淡雅。木栈道沿海边峭壁的山势而建，行走其上便可欣赏山崖顶端的景观。建在晶石地质和霍克斯伯里砂岩上的木栈道，走势高低曲折，与山石紧密吻合，将五个形态各异的观景点串联起来。

木栈道上的木材与网眼型的玻璃纤维格子对接，让水和阳光可以透进来，下面是在零星各处的绿色植被。建筑材料的选择主要考虑到质量、耐久性和可持续发展性。一定数量的围栏在游客感受美景的同时保证了他们的安全。

鉴于建筑地点的情况比较复杂，设计师需要做严格的岩土工程技术计划，以解决当地露出地面的岩层和侵入岩墙的问题。一些地方甚至需要打地桩来维持不甚稳固的填土斜坡。敏感的环境意味着施工通道只能局限在四米范围以内。这项工程最初为了保护维沃里公墓这一最具悉尼历史特征的壮观历史古迹，因为一年70万人的客流量使公墓不断地遭到损坏。

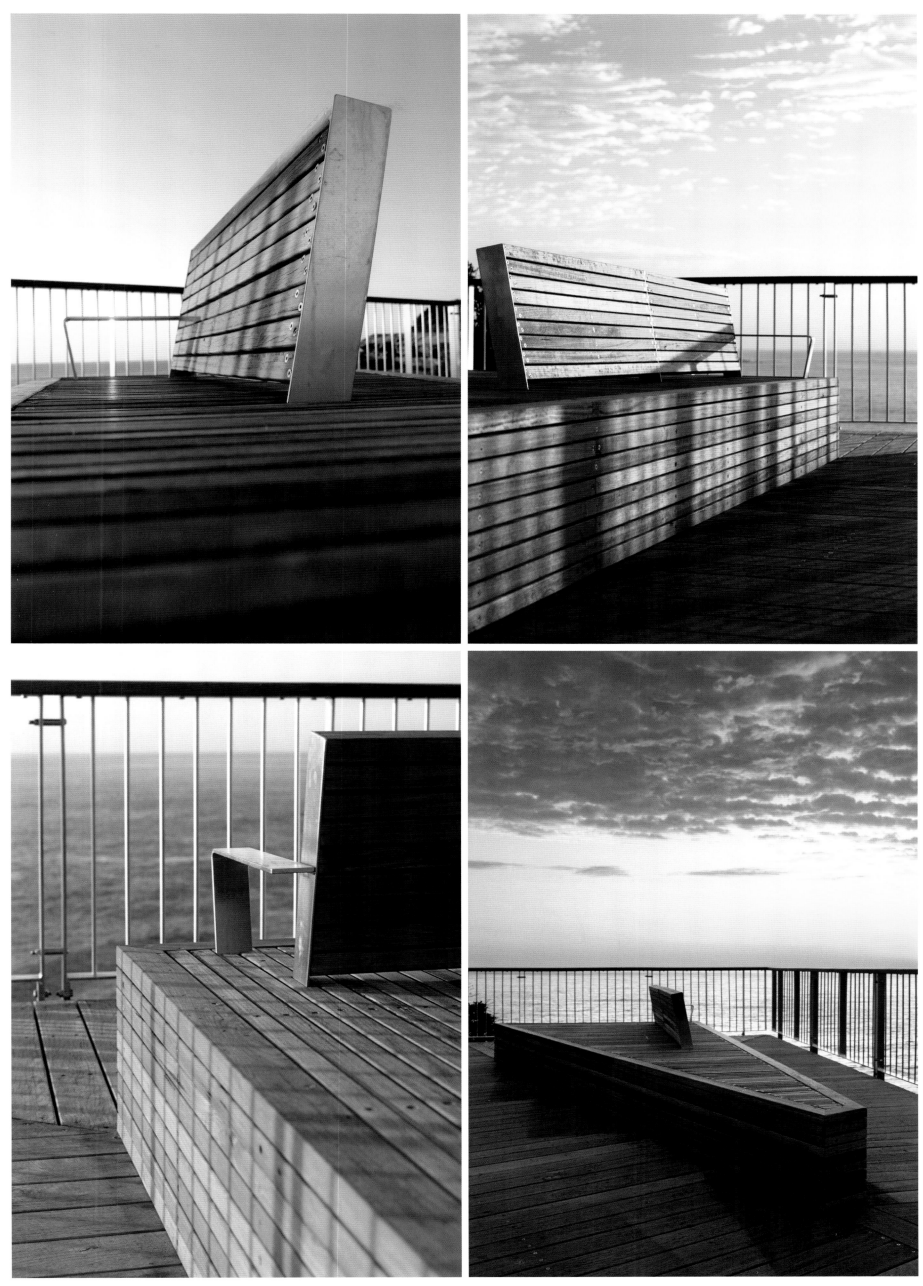

LANDSCAPE ARCHITECT HASSELL **COLLABORATOR** Northern Territory Government / Darwin Waterfront Corporation / AMP Capital / Toga Group

DARWIN WATERFRONT PUBLIC DOMAIN

/ NORTHERN TERRITORY, AUSTRALIA

达尔文海滨公共区域

The vision for the Darwin Waterfront project was to transform the city's redundant industrial port facility into a world class, mixed-use urban community and public haven, attracting residents, business travellers and tourists alike. Creating a new harbour front precinct, linking the water to the city, and revitalising the derelict but historically important site was the challenge for the new development.

Designed as a series of staged precincts, this first public domain precinct set the precedent for all future stages. It implemented many of the critical principles and strategies proposed in the master plan to reconnect the harbour to the broader urban fabric, incorporate the site's historic headlands, re-establish natural vegetation and reflect Darwin's strong cultural heritage.

The public domain comprises structured and identifiable recreational spaces. It is contained by residential buildings, an apartment hotel and the Darwin Convention Centre, and a variety of retail and dining opportunities which contribute to the activation of the precinct.

A waterside connection was critical in establishing a beachfront setting for the tropical city. The new swimming beach and the wave lagoon are the heart of the waterfront. An active beach defined by patrolled paddling pools and a contained sand beach have been established for year-round swimming. Additionally, a natural amphitheatre and open parklands provide relaxed, flexible outdoor space for recreation and events.

The project demonstrates how integrated public realm design can remediate, reactivate and bring new life to previously industrial areas of a city. Darwin Waterfront has evolved into a distinctive entertainment, cultural and residential precinct that enhances the desirability of Darwin as a place to live and visit.

CLIENT Macmahon   **AREA** 25,000 m²   **PHOTOGRAPHER** Brett Boardman / David Silva

　　达尔文滨水项目的愿景是，将多余的工业港口设施改造成世界级的综合功能城市社区和公众天堂，以吸引当地居民、商务差旅和旅游人员。新开发项目的挑战在于打造新的港口滨水片区，将水域与城市相连，为遭到荒废但仍具重要历史价值的基地重塑活力。

　　首个公共空间片区设计为一系列分阶建设的部分，并为所有后续阶段提供前提。其中贯彻了在总体规划中提出的诸多重要原则和策略，从而将港口与更大范围的城市肌理连接起来，将基地的历史性海岬结合其中，重造自然植被，反映达尔文重要的文化遗产。

　　公共空间由结构性的醒目的休闲空间构成，并由住宅建筑、酒店公寓、达尔文会议中心，以及各类零售与餐饮空间封闭而成，这些部分共同促进了片区的活跃氛围。

　　滨水联系的重要性在于为热带城市建立起海滩环境。新建的游泳海滩和造浪泻湖作为滨水区的中心。配备了巡查员的划船水塘界定出的活跃的海滩以及一片封闭沙滩都已经建造，全年可用于游泳活动。此外，自然露天场地和开放绿地提供让人放松而灵活的室外空间，可用于休闲与大型活动。

　　项目展现综合公共空间设计如何可以修补并重新活跃从前的城市工业区，并为其注入新的生机。达尔文滨水区已演变为富有特色的娱乐、文化和住宅区，进一步体现了达尔文成为居住地和旅游目的地的价值所在。

/ 021

LANDSCAPE ARCHITECT ASPECT Studios   TEAM ASPECT Studios / Tweed Shire Council

JACK EVANS BOAT HARBOUR - TWEED HEADS - STAGE 1
/ NSW, AUSTRALIA

杰克·埃文斯船港——堤维德岬一期工程

CLIENT Tweed Shire Council AREA 49,000 m² PHOTOGRAPHER Simon Wood

The design of Jack Evans Boat Harbour reveals the singular beauty of the ever-shifting inter-tidal zone as an inhabitable landscape.

The primary organisational element of the design is a simple, stepped, concrete gesture that frames the harbour edge. The design's success lies in its creation of a place of recreation; whilst allowing the tidal nature of the river itself to create a unique and ever-changing experience of the parklands. It is this simple solution which will ensure the longevity of the Jack Evans Boat Harbour parklands as they are placed under the pressures of a growing coastal population and the many pressures of urban densification.

A series of distinct relationships with the water develop along the harbour edge – a new beach and beach deck, rocky headland, an "urban pier", boardwalk, water amphitheatre, swimming areas, fishing points and opportunities for water craft, all of which are designed to withstand frequent tidal and storm surge inundation, and to "future proof" the surrounding parklands against the effects of climate change and sea level rise. In addition, the reshaping of the shoreline has enabled the development of an "all abilities" access ramp to the water at all tidal levels, a unique recreational opportunity for the area.

杰克·埃文斯船港景观设计项目是一处宜居景观,显示出潮间带不断变化所带来的非凡美景。

该设计的基本组织元素是简单的、阶梯式的混凝土造型。它们共同围合出船港的边缘。此次设计的成功之处在于创建了休闲娱乐场所,同时利用河流本身的潮汐特性创造出独一无二、不断变化的公用绿地体验。海岸人口不断增长,城市越来越拥挤,使杰克·埃文斯船港的绿地承受着巨大的使用压力,而正是这种简单的设计方案,将确保其长期存在。

沿着港口开发出一系列与水域密切相关的项目,其中包括:新海滩和海滩甲板、岩石海岬、"都市码头"、木板路、圆形水剧场、游泳场、钓鱼处和划船处。这些项目的设计旨在抵抗频发的潮汐和暴风雨所带来的洪水泛滥,并使公园的周边地区抵抗未来气候变化和海平面上涨所带来的影响。此外,海岸线改造后形成了一个在任何潮位段都"全能"的斜坡接岸,因此,这里也是该地区独一无二的娱乐休闲场所。

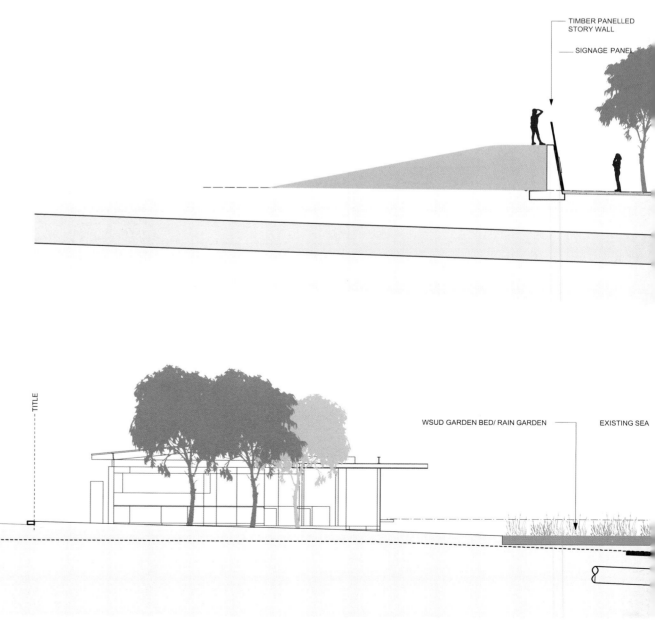

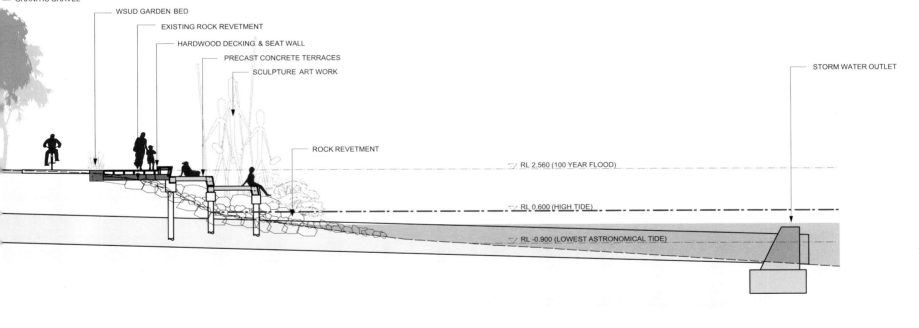

- GRANITIC GRAVEL
- WSUD GARDEN BED
- EXISTING ROCK REVETMENT
- HARDWOOD DECKING & SEAT WALL
- PRECAST CONCRETE TERRACES
- SCULPTURE ART WORK
- ROCK REVETMENT
- STORM WATER OUTLET

RL 2.560 (100 YEAR FLOOD)
RL 0.600 (HIGH TIDE)
RL -0.900 (LOWEST ASTRONOMICAL TIDE)

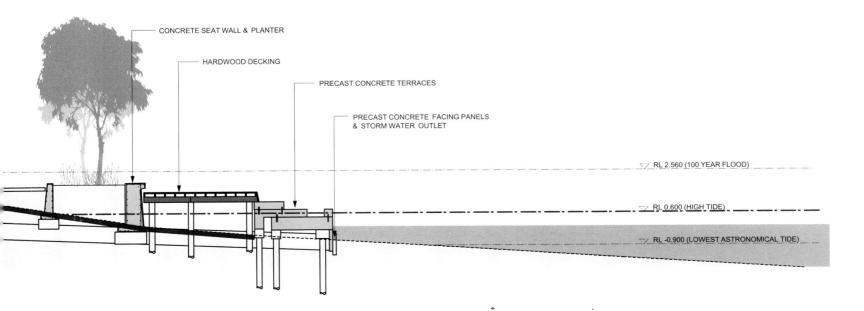

- CONCRETE SEAT WALL & PLANTER
- HARDWOOD DECKING
- PRECAST CONCRETE TERRACES
- PRECAST CONCRETE FACING PANELS & STORM WATER OUTLET

RL 2.560 (100 YEAR FLOOD)
RL 0.600 (HIGH TIDE)
RL -0.900 (LOWEST ASTRONOMICAL TIDE)

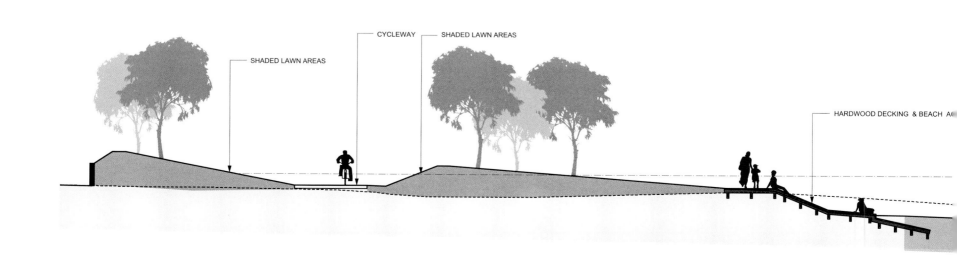

SHADED LAWN AREAS · CYCLEWAY · SHADED LAWN AREAS · HARDWOOD DECKING & BEACH A

EXISTING SEA BEES

RL 2.560 (100 YEAR FLOOD)

RL 0.600 (HIGH TIDE)

RL -0.900 (LOWEST ASTRONOMICAL TIDE)

LANDSCAPE ARCHITECT Vicente Guallart / Maria Diaz   **FIRM** Guallart Architects

KEELUNG MARITIME PLAZA
/ TAIWAN, CHINA

基隆海洋广场

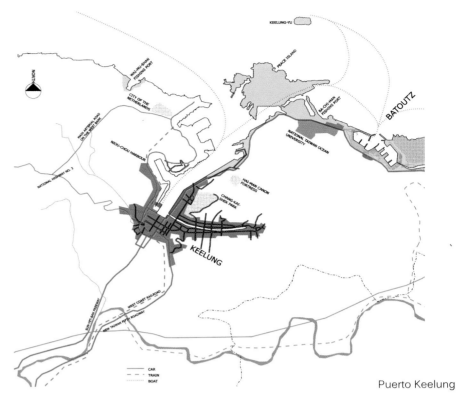

Puerto Keelung

PHOTOGRAPHER Adriá Goula

Keelung is the port of Taipei, Taiwan. Located 30 km to the north of Taipei, it is one of the most important container ports in Asia. Keelung has all the vitality of a major port, with one of the most bustling night-time markets in the Far East and an extensive and multifarious central commercial area adjoining the port. The city nevertheless bears the traces of rapid economic growth. Its principal transportation infrastructures —roads, railway lines, and the port itself — continue to limit the creation of quality public spaces in the downtown area.

In the light of this, the authorities invited projects as part of the plan to create new "Gateways" in Taiwan, oriented toward defining the interaction between the port and the city. In fact, the fundamental issue to be resolved by the various projects drawn up during the different phases of the competition and in the subsequent construction scheme was how to identify the characteristics of a new central public space for the city with which the citizens of Keelung could identify. Historically, Asian cities have a strong tradition of the use of public space and a dynamic inside-outside relationship that has generated numerous instances of cities, neighbourhoods and residential or commercial sectors of great urbanity. However, the economic development of recent years seems to have oriented urban development toward public spaces more in line with the American model, based on the habitability of air-conditioned interior spaces or urban mobility based on the car that makes the car park one of the fundamental interchanges in urban life. This makes it difficult to identify significant urban spaces created in the last few years that respond to the traditional dynamic occupation of the public space. Keelung is seeing the start of another process that is already present in most American, European and Australian cities, in which the port-city interaction is redefined in the interests of a greater public use of port spaces. In this way the historic port zones, which are normally in the proximity of central urban places, are ceded by the port to the city as a site for leisure and commercial uses, sports ports and even hotel and residential zones. Darling Harbour in Sydney, the Port of Boston or Port Vell in Barcelona are examples of such transformations.

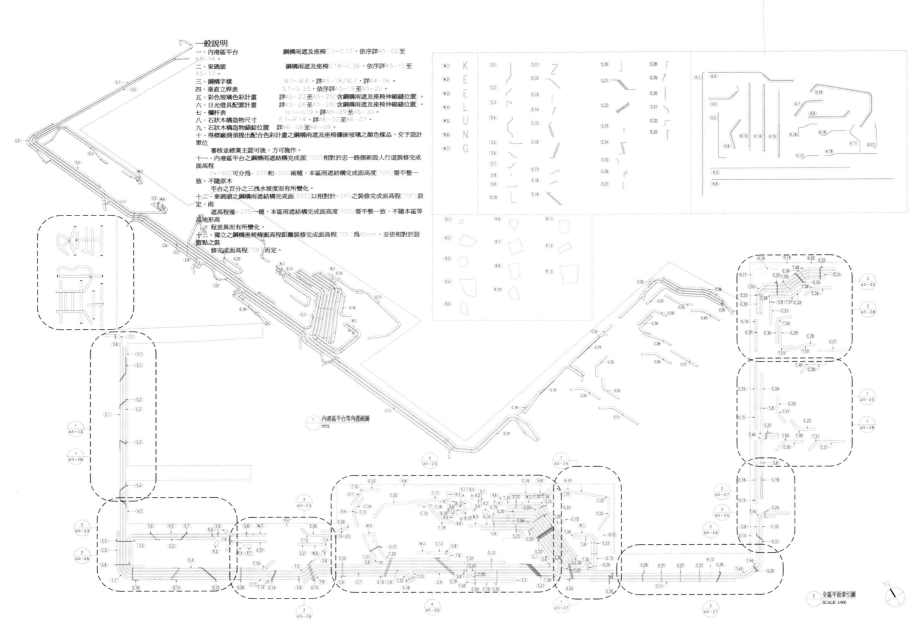

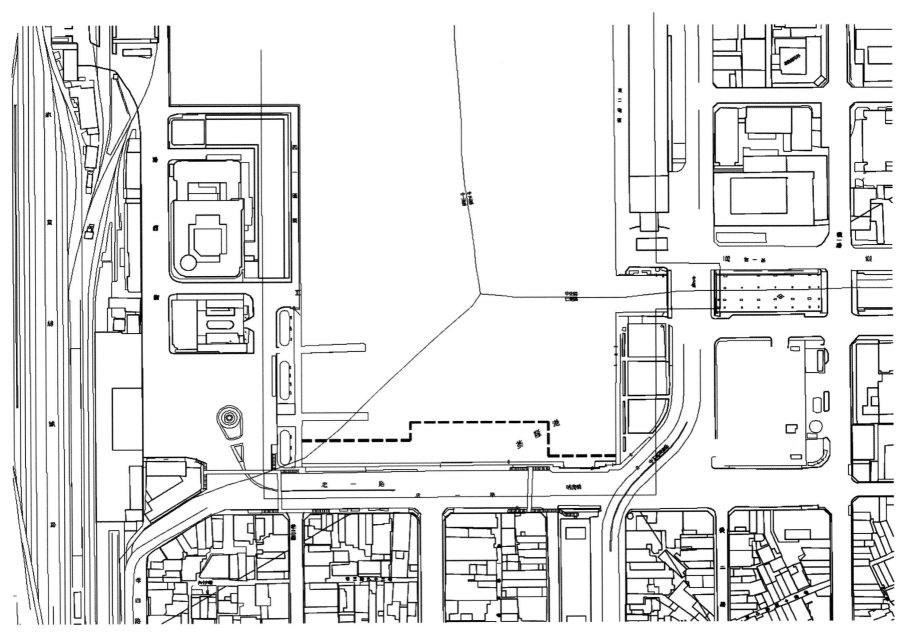

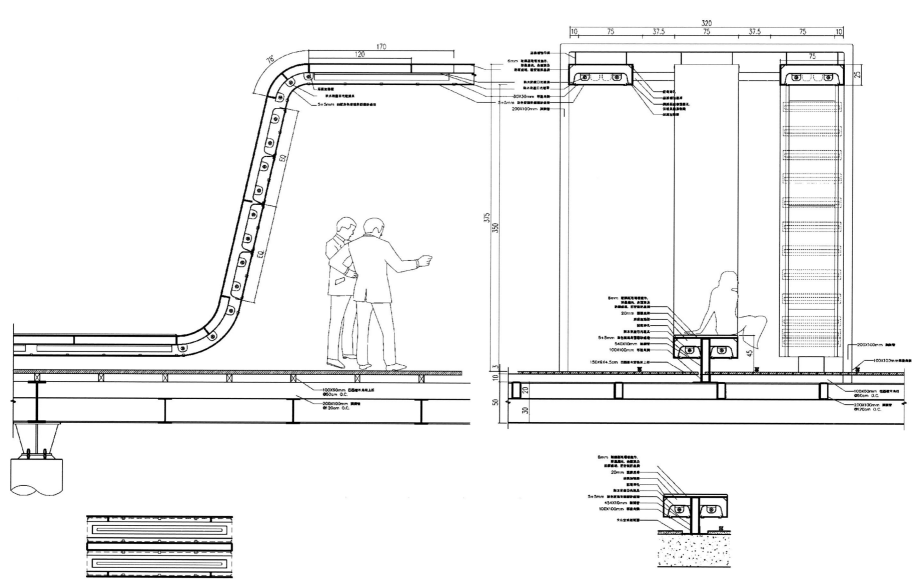

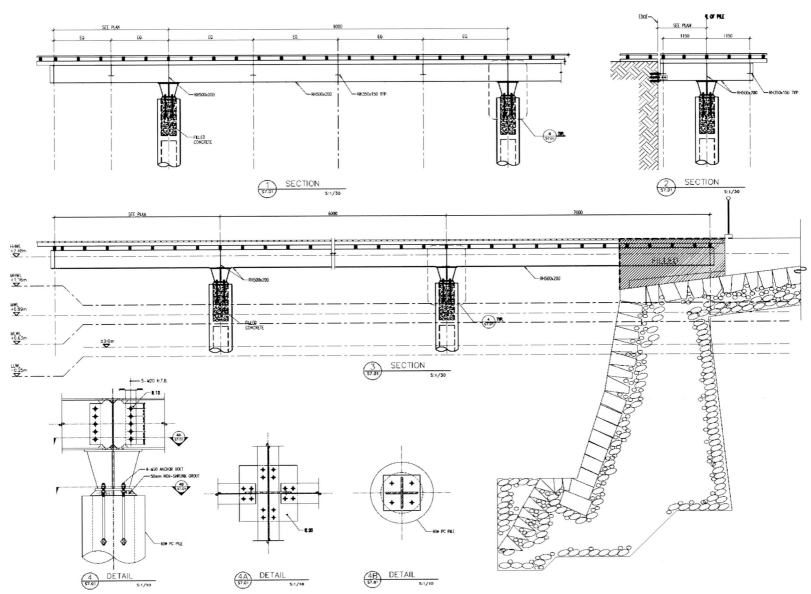

/ 045

　　基隆港是台湾岛的一座海港，位于台北以北 30 千米处，是亚洲最重要的集装箱港口之一。基隆拥有成为一座主要港口的所有潜质：这里有远东最热闹的夜市之一，并且与一块广阔而繁华的中央商业区毗邻。然而，快速的经济增长给城市留下了痕迹。城市的主要交通基础设施——道路、铁路专用线和港口本身——都一直限制着市中心优质公共空间的发展。

　　鉴于此，当地政府将该项目作为创建台湾新"门户"计划的一部分，使其以港口与城市之间的互动为导向。事实上，在设计方案竞标不同阶段完成的各种项目设计方案以及后期的施工方案里，亟待解决的基本问题是：如何确立新的城市中央公共空间的特征，并使该特征为基隆市民所接受。从历史上看，亚洲的城市在建造景观时，有一种根深蒂固的传统方式，即利用公共空间以及富有活力的内外关系。许多城市、居民区、住宅或带有浓厚都市气息的商业区都是在这样的传统方式下建造而成。但是，基于装有空调设备的室内空间的宜居性，或鉴于汽车使得停车场成为都市生活中重要的"换乘"空间这一事实所带来的都市流动性，近年来的经济发展似乎将都市开发更多地定位在效仿美国模式的公共空间上。上述定位使人们难以识别近年来所创建的、符合传统公共空间设计动感理念的重要都市空间。基隆正见证着另一种开发进程，而这些进程已经在美国、欧洲和澳大利亚的大多数城市启动；在这一进程中，港口与城市的交互关系意味着让更多的市民享受港口公共空间。以这种方式，通常位于城市中心附近的历史港区就由港口划归给城市，并作为休闲区、商业区、体育运动区、酒店和住宅用地。悉尼的达令港、波士顿港以及巴塞罗那的威尔港，都是这种转变下的例子。

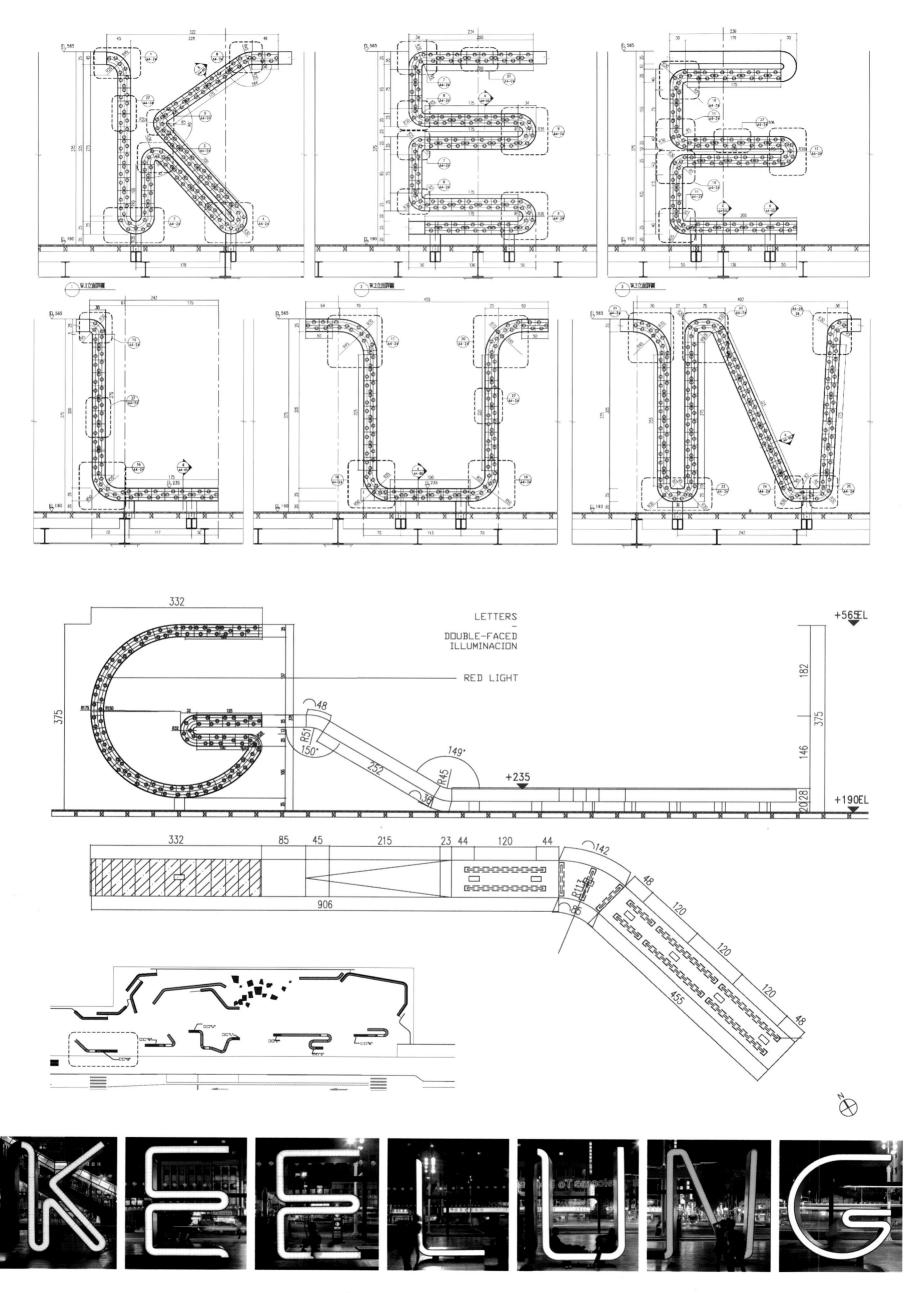

LANDSCAPE ARCHITECT Claude Cormier + Associés inc.　　**CONSULTANT ENGINEERING** Les services EXP inc.　　**LIGHTING DESIGNER** Éclairage Public

CLOCK TOWER BEACH
/ QUEBEC, CANADA

钟楼海滩

The introduction of an urban beach to the Quai de l'Horloge is an ideal addition to the recreational and cultural redevelopment project at Montreal's Old Port. A stroll along the quay opens a stunning panorama to the visitor: the mighty St. Lawrence River, the impressive Jacques Cartier Bridge, Calder's iconic sculpture, and picturesque Old Montreal as a backdrop to it all.

The project consists of two closely-linked components. The first is the creation of an urban beach at Point de l'Horloge, with its elegant clock tower built in 1921, and along the lower quay bordering the marina. A huge stairway-ramp makes this convivial venue accessible to all, offering a new and novel approach to city living. Beach umbrellas and weeping willows, brightly coloured chairs and fixtures, showers and mist stations, a boardwalk and silky sand all combine to offer visitors a few moments of sheer idleness in a breathtaking setting.

The second component is the parking area, clearly defined by rows of trees that reproduce the triangular layout of the quay. Fed by surface water run-off, the larches, willows, shrub beds, and perennials add cool green ambiance to the space. At the far west end of the site, shooting up from a mound around which a roundabout loops, hundreds of sticks in three shades of blue generate an intriguing pixilation effect. This "porcupine" installation revives a conceptual element emblematic of the firm.

MAIN MATERIALS Umbrellas / Adirondack compact chair / sand / misters   **AREA** 13,000 m²   **PHOTOGRAPHER** Claude Cormier + Associés inc. / Marc Cramer / Gilles Arpin

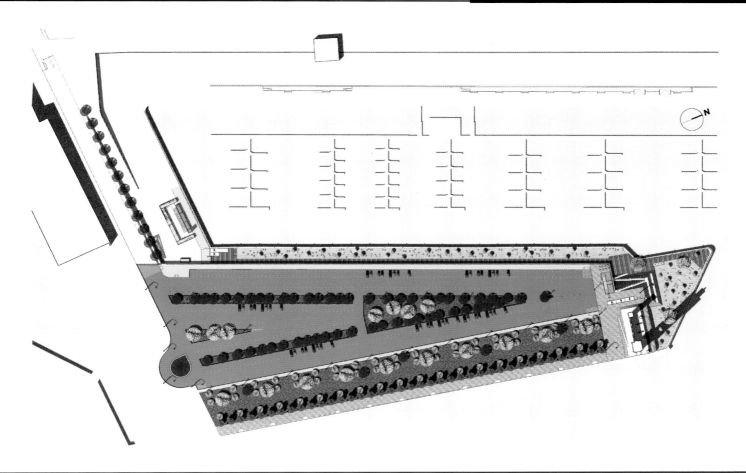

　　钟楼码头新建的城市滨海区，无论在文化上还是在休闲娱乐上，都给蒙特利尔老港的重建工程增色不少。沿着码头漫步，游客可以欣赏到美轮美奂的全景：壮丽的圣劳伦斯河、雄伟的雅克·卡地亚大桥、考尔德的标志性雕塑以及映衬着这一切的风景如画的蒙特利尔老城区。

　　这一工程由紧密联系的两部分组成。第一部分是在钟楼所在地建一个城市滨海区。这里有建于1921年的优雅的钟楼和沿着小艇停靠区的低地码头。巨大的斜坡台阶便人们到达这欢乐的场所，为城市生活增添了新的乐趣。海滨遮阳伞、随风摆动的垂柳、色彩明亮的桌椅、淋浴点和喷雾站、绵延的木板路和细软的沙滩，这一切交织在一起，使游客享受片刻的悠闲，犹如进入了美丽的仙境。

　　工程的第二部分是新建一个停车场。三角形的停车场边缘整齐地种植着成排的树木，与码头三角形的布局设计一致。落叶松、垂柳、灌木植被和多年生植物在地表径流的滋养下茁壮成长，为此地增添了些许绿色气息，也为人们提供了片片阴凉。在景点最西边环状公路围绕的筑堤上，数以百计的蓝色棒作喷涌而出状，三种明暗色调不同的蓝色营造出一种3D动画效果。这个形似豪猪的建筑使人想起象征着公司的概念元素。

LANDSCAPE ARCHITECT Mayslits Kassif Architects **TEAM** Ganit Mayslits Kassif / Udi Kassif / Oren Ben Avraham / Galila Yavin / Michal Ilan / Maor Roytman

TEL AVIV PORT PUBLIC SPACE REGENERATION
/ TEL AVIV, ISRAEL
特拉维夫港口公共空间的改造

| **CLIENT** Marine Trust Ltd., | **AREA** 55,000 m² | **PHOTOGRAPHER** Adi Brande / Galia Kronfeld / Daniela Orvin / Albi Serfaty |

Situated on one of Israel's most breathtaking waterfronts, the Tel Aviv Port was plagued with neglect since 1965, when its primary use as an operational docking port was abandoned. The recently completed public space development project by Mayslits Kassif Architects, managed to restore this unique part of the city, and turn it into a prominent, vivacious urban landmark.

The architects viewed the project as a unique opportunity to construct a public space which challenges the common contrast between private and public development, and suggests a new agenda of hospitality for collective open spaces. The design, a winner of an open competition held in 2003 (entry submitted by Mayslits Kassif Architects in collaboration with Galila Yavin) was quickly brought to life by a new management, with locals and visitors flocking to the revamped port even before the project was completed.

The design introduces an extensive undulating, non-hierarchical surface, that acts both as a reflection of the mythological dunes on which the port was built, and as an open invitation to free interpretations and unstructured activities. Various public and social initiatives – from spontaneous rallies to artistic endeavors and public acts of solidarity – are now drawn to this unique urban platform, indicating the project's success in reinventing the port as a vibrant public sphere.

Nowadays when approximately 2.5 million people visit the Tel Aviv Port every year – a record number for a metropolitan area spanning 1 million residents, in a country of 7 million – the port's public spaces renewal is considered one of the most influential projects of its kind in Tel Aviv. Alongside receiving international recognition and several prestigious architectural awards, such as the Rosa Barba European Landscape Prize for 2010, it receives great affection from the public and is ranked as the most beloved recreation space by the inhabitants of Tel Aviv's metropolitan area. Being a new urban landmark which revives the city's waterfront, the project became a trigger for a series of public space projects along Tel Aviv's shoreline which altogether revolutionize the city's connection to its waterfront.

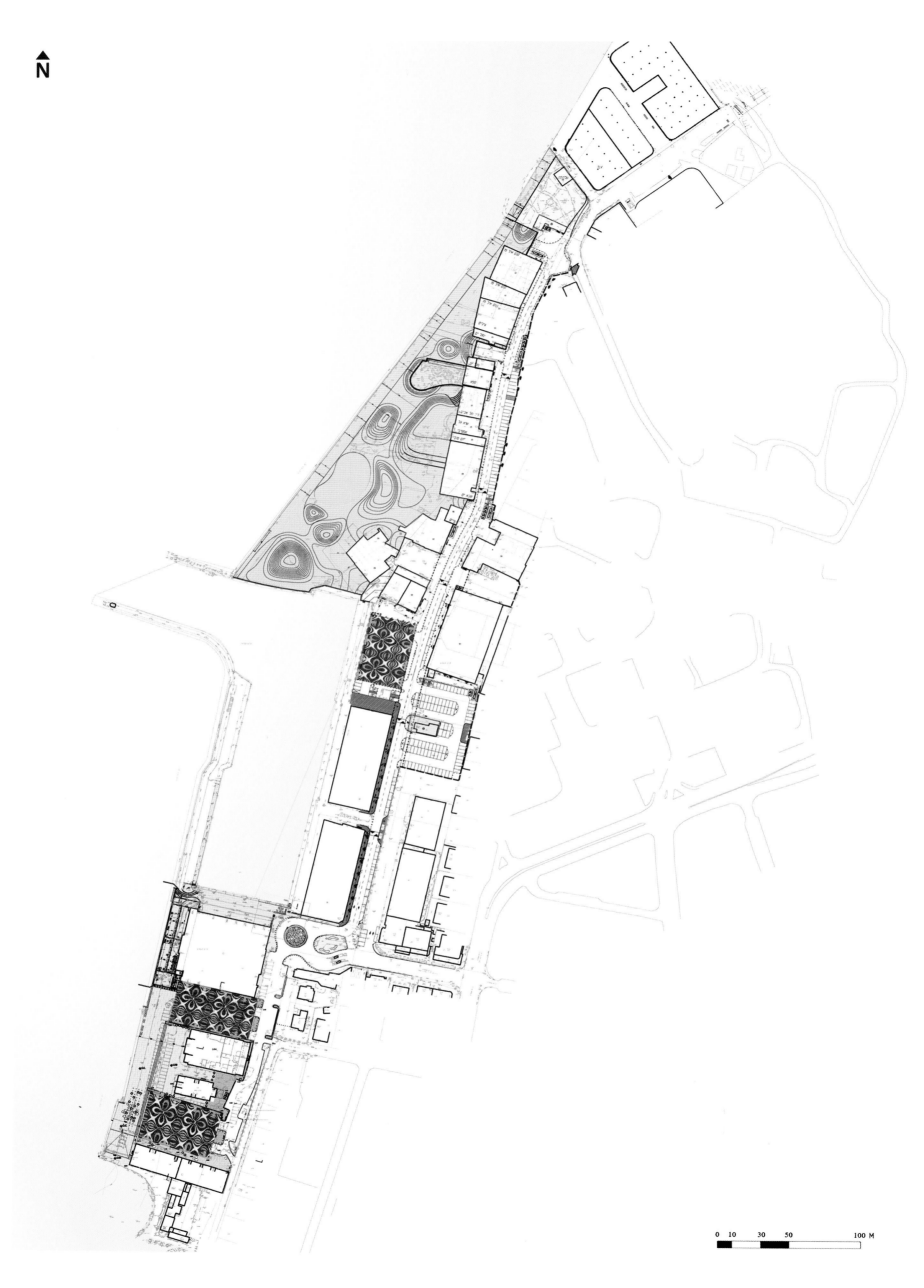

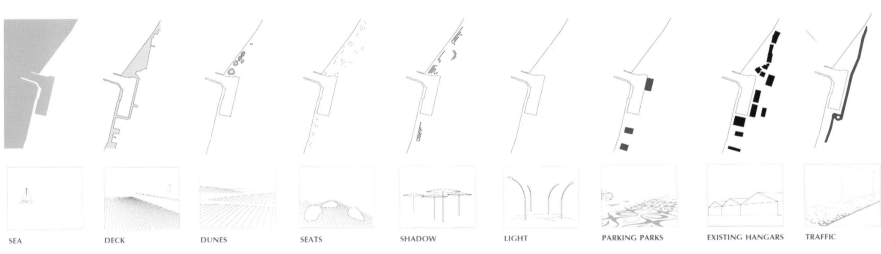

| SEA | DECK | DUNES | SEATS | SHADOW | LIGHT | PARKING PARKS | EXISTING HANGARS | TRAFFIC |

　　特拉维夫港口位于风光旖旎的以色列滨水之畔，原是一座货运港口，但自 1965 年停用之后，便淡出了人们的视野。最近由 Mayslits Kassif 建筑事务所完成的公共空间开发项目，力图重建特拉维夫市这一独特的区域，并将它打造成杰出而富有活力的城市地标。

　　建筑师们把这个项目视为塑造公共空间的一个千载难逢的机会，挑战普遍为人接受的私人开发项目及公共土地开发项目对立的观念，提出公共露天空间要环境亲民的新主张。这个设计在 2003 年的公开竞赛中脱颖而出（由 Mayslits Kassif 建筑事务所与 Galila Yavin 合作），并通过全新的项目运作，很快得以实施。在此项目尚未完全竣工时，当地居民和游客就已成群地涌向这个翻建的港口去一睹它的风采。

　　项目建造了一处开阔、波状起伏的场地，不仅是港口原有的神奇沙丘的缩影，也是进行各种自由表演或举行各种非正式的活动的最佳场所。多样化的公共和社会活动——即

兴集会、艺术活动、团体演出——都被纳入这个独特的城市舞台，这表明项目已成功地将港口改造成一个生机勃勃的公共空间。

如今，每年大约有250万人参观特拉维夫港，这对人口100万的特拉维夫市，乃至全国人口700万的以色列来说，都是个历史纪录。港口公共空间的改造被认为是特拉维夫市此类项目中最具影响力的项目之一。特拉维夫港深受公众的喜爱，它被列为特拉维夫都市区居民最喜爱的休闲娱乐场所。除此之外，它还获得了国际的认可和一系列颇有声望的建筑学奖项，如2010年度罗莎·芭芭欧洲景观奖。作为复兴城市水域的新城市地标，此项目触发了特拉维夫海岸区域一系列公共空间项目的展开，由此带来城市与水域关系的变革。

| LANDSCAPE ARCHITECT FoRM Associates | AREA 5,400 m² | PHOTOGRAPHER FoRM Associates |

TRAFFORD WHARF PROMENADE
/ MANCHESTER, UK
特拉福德滨河长廊

Quayside designed by London-based FoRM Associates in front of Imperial War Museum North by Daniel Libeskind is now open to the public.

The new quayside completes the first section of the Irwell River Park masterplan, 8km long river edge park linking Salford, Manchester and Trafford developed by FoRM Associates in 2010. With the adjoining new Media City foot bridge by Wilkinson Eyre Architects the quayside delivers an important new strategic circulation loop in the Quays, a key regeneration zone in Greater Manchester. The loop helps to transform the experience of walking in the area through linking Media City UK - the new home of the BBC, with the IWMN, Manchester United Stadium and Lowry Arts Centre.

FoRM's design of the quayside plays with convex and concave geometries, creating an imaginative public realm complementing the designs of both the IWMN and the new Media Bridge. Importantly the project also actively orientates the IWMN towards its water edge with its newly constructed additional entrance now admitting over 50% of the museums visitors. The original severance to movement along the Manchester Ship Canal was resolved by a newly constructed deck and stepped area built over water, delivering generous pedestrian and cycling connectivity as well as a series of public realm spaces that serve as informal performance/educational resource for the museum. Terraced seating leading down to the water's edge provides excellent views of the area making the new quayside already a popular destination both during the day and night.

"FoRM Associates were not known to Peel before the Trafford Wharf commission. They proved to be one of the most innovative and professional consultants I have worked with. The project was delivered not just on time and entirely within a tight budget but with great flair and imagination. FoRM clearly understood their brief and delivered not just a visual delight but, in concert with the structural engineer produced a technically sound and simply constructed solution" Ed Burrows, Property Director, Peel Media said.

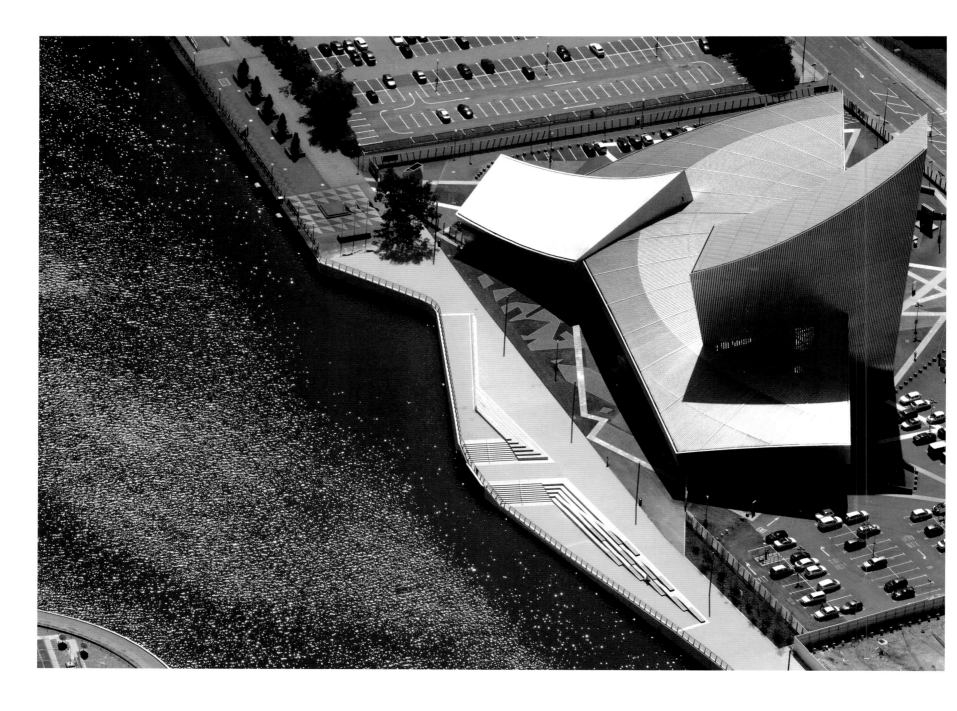
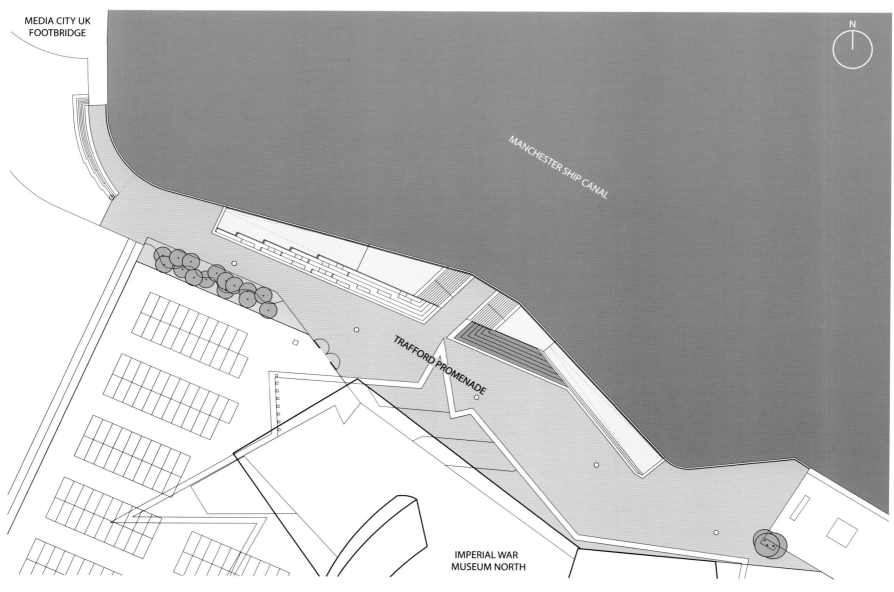

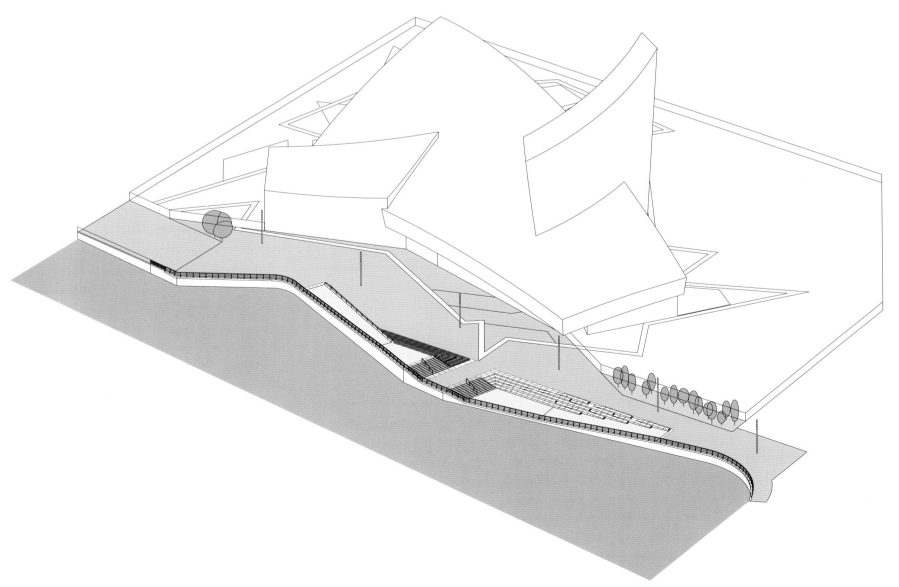

该码头项目位于丹尼尔·里伯斯金设计的曼彻斯特帝国战争博物馆北馆的前方，由总部设在伦敦的 FoRM Associates 负责设计，现已对公众开放。

8千米长的河畔公园连接了索尔福德、曼彻斯特和特拉福德，它是由 FoRM Associates 于 2010 年开发的项目。新的码头广场完成了艾尔韦尔河公园总体规划设计的第一期。码头广场连接了由威尔森·艾尔建筑师事务所设计的新媒体城步行天桥，在码头区域内形成一条全新又重要的策略性环形路线，构成了大曼彻斯特地区主要的重建区域。这个环形路线有助于通过打通英国媒体城——英国广播公司新总部大楼、曼彻斯特帝国战争博物馆北馆以及曼彻斯特联队体育场和洛利艺术中心之间的联系，丰富了该地区人们的步行新体验。

FoRM 对于码头广场周边的设计采用了凹凸结合的几何造型，打造出一个充满想象力的公共空间，与曼彻斯特帝国战争博物馆北馆及新媒体桥的设计形成互补之势。该项目的独特之处在于，灵活地利用新建的另一入口，将曼彻斯特帝国战争博物馆北馆的面水一侧开放，该入口现可接纳超过一半的博物馆游客进入。过去，沿曼彻斯特运河的交通阻滞；如今，新建的桥面及水上阶梯使宽阔的步行道和自行车道得以延伸，也为博物馆提供了一系列用于非正式演出／教育资源服务的公共空间。阶梯状座椅沿坡一直延伸到水边，为游客提供了绝佳的视野，使码头广场无论在白天还是黑夜都成为游客喜爱的旅游之地。

Peel 媒体的主管艾德·巴罗斯说道："FoRM Associates 是在特拉福德滨河长廊项目之后才被 Peel 媒体所熟知，实践证明他们是我合作过的最具创新意识和最为专业的顾问团之一。该项目工期紧，预算紧，但都按时交付，且极具创意和想象力。FoRM 很清楚他们的宗旨，在创造出建筑视觉盛宴的同时，与结构工程师通力合作，提出技术合理、构造简洁的解决方案。"

LANDSCAPE ARCHITECT Place Design Group **CLIENT** Sunshine Coast Regional Council

BULCOCK BEACH FORESHORE REDEVELOPMENT

/ AUSTRALIA

布卡克海滩重建项目

PLACE Design Group was commissioned as lead consultant in 2007 by the Sunshine Coast Regional Council to undertake master plan review, design development, stakeholder consultation and to prepare detailed tender documentation for the Bulcock Beach Esplanade Redevelopment project. The procurement of various environmental and building approvals from Local, State and Federal Agencies was also required. These approvals were facilitated by proactive discussions with the various agencies and the preparation of an Environmental Management Plan Framework greatly assisted this complex process.

The Bulcock Beach foreshore is a unique site with significant environmental, recreational and scenic values. The project also involved challenges associated with infrastructure renewal and upgrade in such a sensitive location and within a limited construction budget. PLACE Design Group coordinated and managed a comprehensive team of sub consultants including structural engineer, quantity surveyor, geotechnical engineers, soil scientist, architect, sculptural artist, interpretive, access, hydraulic and electrical consultants through the design and construction phases of the project.

PLACE Design Group also introduced Street & Garden Furniture, Chelmstone, Bell Stainless Steel and Urban Art Projects to the project and worked with these specialist suppliers and fabricators to achieve an outstanding and robust place making outcome. The design, furnishings, detailing and materials selection is both responsive to the site and its Pumicestone Passage marine location and will have enduring value to the community. The redevelopment has revitalized a popular destination for locals and tourists providing greatly improved beach access and public recreational opportunities.

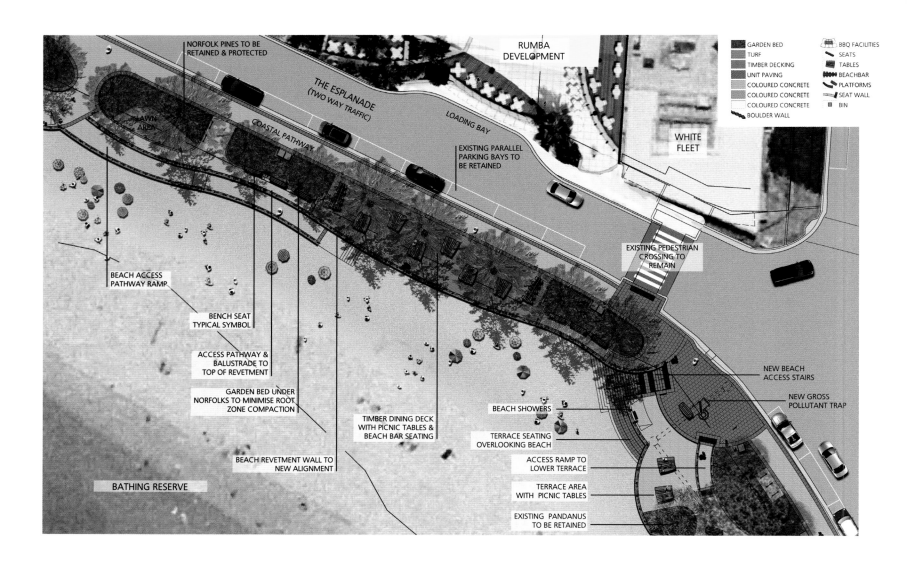

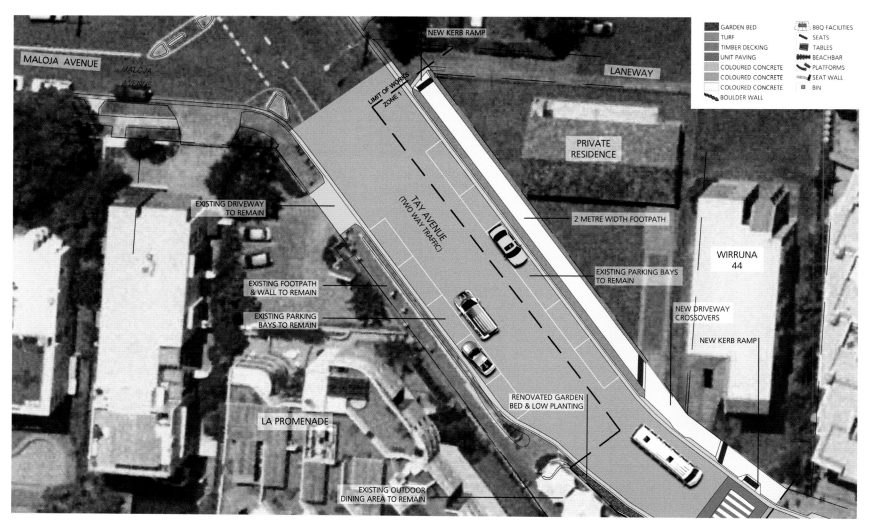

2007年，阳光海岸地区议会委任PLACE设计集团为布卡克海滩滨海路重建项目的首席顾问，承担项目的总体规划评议、设计开发、股东咨询三方面的工作，并为项目准备详细的招标文件。该项目需从当地、所在州和联邦政府的有关部门获得各种环境和建筑上的许可。PLACE设计集团与各部门进行积极主动的商讨，促进了许可的取得，《环境管理计划框架》的草拟也极大地推动了这一复杂的进程。

布卡克海滩的前滩是一个独特的景点，具有重要的环境、娱乐和风景价值。在地理位置敏感、建设预算经费有限的条件下，该项目面临着基础设施更新和升级的挑战。在项目的设计和施工阶段，PLACE设计集团始终对咨询团队进行协调和管理。这一综合团队成员包括结构工程师、工料估算师、岩土工程师、土壤学家、建筑师、雕刻艺术家、解说顾问、宣传顾问、液压和电气顾问。

PLACE设计集团也将街园家具城、Chelmstone公司、贝尔不锈钢公司和都市艺术工程公司引进到该项目中，并与这几家专业供应商和制造商合作，使这一建筑景观既出类拔萃又稳固耐用。该项目在设计、装饰、细节和材料的选择上，考虑了景点功能及其浮石通道的海上位置，将产生持久的社会价值。通过极大地改善海滩通道和增加市民休闲娱乐的机会，该重建项目使布卡克海滨重新恢复了活力，为当地居民和旅游者提供了一个备受欢迎的休闲娱乐圣地。

CALOUNDRA FORESHORE PROJECT
CONCEPT DESIGN - BENCH SEAT BACKREST + ARM

CALOUNDRA FORESHORE PROJECT
CONCEPT DESIGN - PICNIC SETTING

CALOUNDRA FORESHORE PROJECT
CONCEPT DESIGN - BACKLESS PLATFORM BENCH

/ 079

/ 081

LANDSCAPE ARCHITECT Blackwell & Associates   **CLIENT** City of Stirling

SCARBOROUGH BEACH URBAN DESIGN MASTER PLAN STAGES 1 & 4

/ SCARBOROUGH, WESTERN AUSTRALIA

斯卡伯勒海滩城市设计规划1、4阶段

SHADE STRUCTURE Shade Sails Australia **PHOTOGRAPHER** Robert Frith - Acorn

Scarborough Beach has long been a popular destination for tourists and Western Australians alike. However, the beachfront was in decline. The infrastructure was run down and anti-social behaviour had become a major problem. The Scarborough Beach Urban Design Master Plan prepared by Blackwell & Associates and endorsed by the City of Stirling in 2005 outlined a logical sequence of stages for the progressive redevelopment of the Scarborough Beach waterfront area.

Subsequent to the successful completion of the Scarborough Beach Urban Design Master Plan and immediately following the City of Stirling securing the rights to hold the National Surf Life Saving Championships at Scarborough, Blackwell & Associates were engaged to implement the $9M first stage of this project.

The intent was to reinvigorate the urban environment and concentrate community activity in the centrepiece of the Master Plan, a world-class multifunction amphitheatre, planned to become home to all manner of events ranging from small community orientated ceremonies, to state, national and international sporting and entertainment events.

This included: improvement of the beach front and reintroduction of the dune system at the western termination of Scarborough Beach Road; the upgrade of West Coast Highway and the car parking and al-fresco areas along the northern extent of The Esplanade plus the creation of extensive grassed terraces and a major new 1,000 seat, "performance standard" amphitheatre.

The designers worked closely with the sail shade fabricators in a collaborative effort, using 3D modelling software, to design an iconic structure that, on one hand maintained sight lines from the upper terraces and, on the other provided appropriate containment and maximum shade protection during the middle of the day. Since its completion, the amphitheatre has become the City's most frequently used and successful outdoor venue.

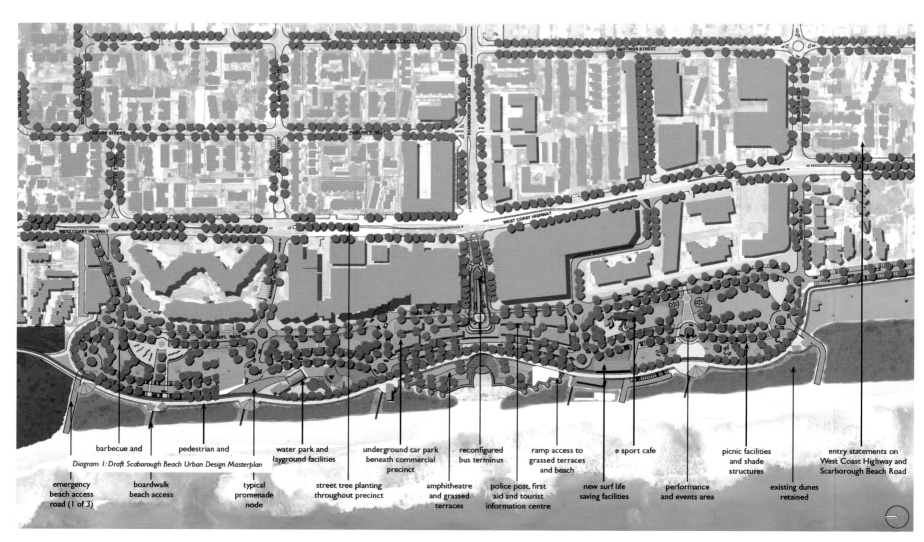

Diagram 1: Draft Scarborough Beach Urban Design Masterplan

- emergency beach access road (1 of 3)
- barbecue and boardwalk beach access
- pedestrian and typical promenade node
- water park and playground facilities
- street tree planting throughout precinct
- underground car park beneath commercial precinct
- amphitheatre and grassed terraces
- reconfigured bus terminus
- police post, first aid and tourist information centre
- ramp access to grassed terraces and beach
- new surf life saving facilities
- e sport cafe
- performance and events area
- picnic facilities and shade structures
- existing dunes retained
- entry statements on West Coast Highway and Scarborough Beach Road

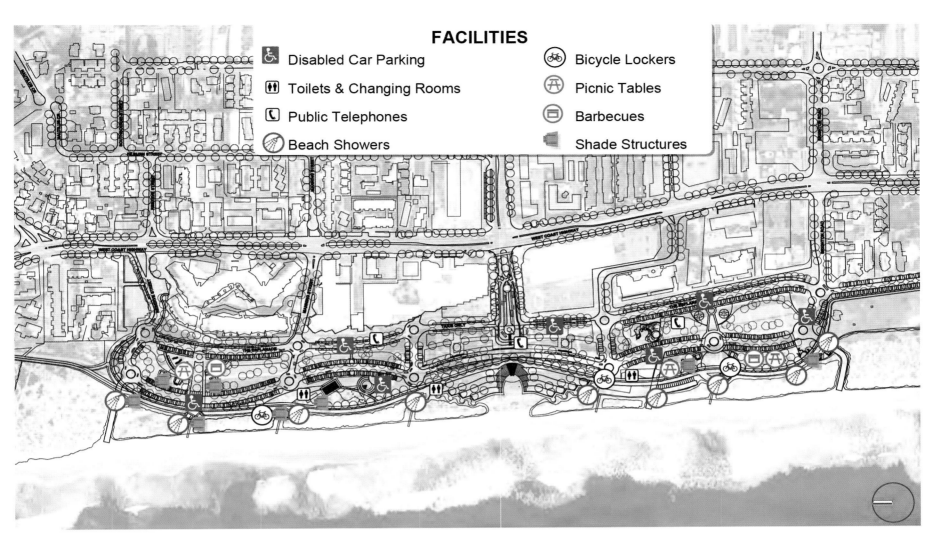

Diagram 12: Facilities Diagram

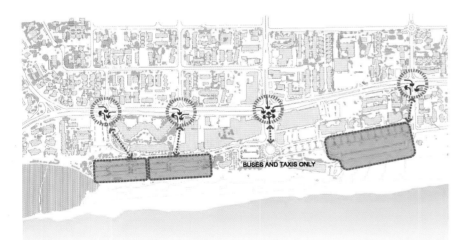

Diagram 7: Broad Access & Car Park Layout Option 1

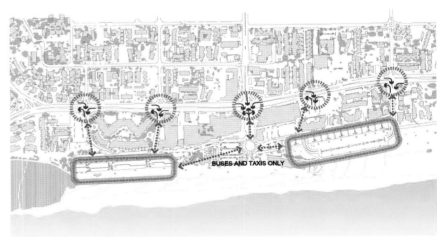

Diagram 8: Broad Access & Car Park Layout Option 2

Diagram 9: Broad Access & Car Park Layout Option 3

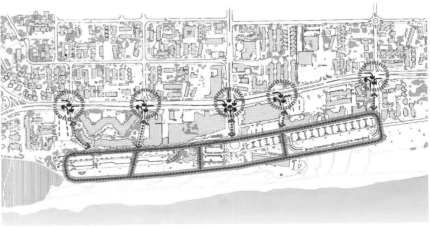

Diagram 10: Broad Access & Car Park Layout Option 4

长久以来,斯卡伯勒海滩都是深受旅游者和西澳大利亚州居民欢迎的旅游胜地。然而,这里日趋败落,基础设施停止运转,反社会行为成为了主要问题。Blackwell & Associates 公司起草了斯卡伯勒海滩城市设计规划,规划中概述了逐步重建斯卡伯勒海滩海滨区不同阶段的前景,并于 2005 年在斯特灵市得以通过。

斯卡伯勒海滩城市设计规划大获成功,斯特灵市获得在斯卡伯勒举办全国冲浪救生锦标赛的资格。之后,Blackwell & Associates 投入到价值 900 万美元的一期工程的建设中。

规划的目的旨在复兴城市环境,并将社区活动集中在总体规划的核心部分,即一个世界级的多功能圆形剧场。在那里可以举行各类活动,从小型的社区典礼仪式到国家级乃至世界级的体育赛事和娱乐休闲活动。

项目内容包括:升级海滨区;在斯卡伯勒海滩路的西面终点站再引入沙丘带;对西海岸公路和停车场以及沿着北部海滨大道的户外区域进行升级;同时,新建广阔的绿草阶梯,加设 1000 个新的座椅,以此来营造符合表演标准的圆形剧场。

设计师与遮阳帆制造商密切合作,利用 3D 立体模型软件来设计标志性结构,一方面保持来自阶梯高处的视线,另一方面提供适当的容量,并且在日间提供最大的遮阳保护设施。圆形剧场建成以来,它已成为城市中使用率最高、最为成功的户外聚集地。

ARCHITECTURE PRACTICE Nyréns Arkitektkontor　　　**LANDSCAPE ARCHITECT** Bengt Isling
TEAM Jacob Almberg / Ronny Brox / Magdalena Franciskovic / Cecilia Jarlöv / Ulrika Lilliehöök / Staffan Malm / Peter Kinnmark Architect MAA

HORNSBERGS STRANDPARK

/ KUNGSHOLMEN, STOCKHOLM, SWEDEN

霍恩博格海滨公园

PHOTOGRAPHER Lennart Johansson / Åke E: son Lindman

Hornsbergs Strandpark is the winner of the Swedish landscape award: Sienapriset 2012.

Hornsbergs Strandpark is where water and land meet in a curvy shoreline and contemporary design, round organic shape and clean lines. It faces west to Ulvsundasjön and the evening sun. The waterfront and the three long floating piers give the visitors a feeling of floating into the light over the water. This is present particularly on hot summer afternoons when the park becomes an oasis for the surrounding residents and used for grilling and swimming. The park features several informal seating areas and a shower with a high seated tank for water heated by the sun that can be used by joggers.

The park is over 700 meters long and consists of four parts. To the west lies a jetty for sunbathing with wooden docks jutting into the lake in different lengths. East of it is Kajparterren formed as a contrast to the organic Strandparken. It is a slightly raised horizontal disc slightly leaning towards the water. Far to the east is an existing part that has been renovated to be more accessible.

The project also includes the Moa Martinson square. For the proposed square design, Nyréns have focused on the spatial situation with a small spot at the edge of Ulvsundasjön and on the artistic adornment associations with the author Moa Martinson. Since the square surface is raised to provide access to the buildings it forms a difference of level with the street. The difference in level consists of a wall and staircase both possible to sit on. The stairs open to the square that is turned diagonally out to the lake with Kajparterren in the foreground.

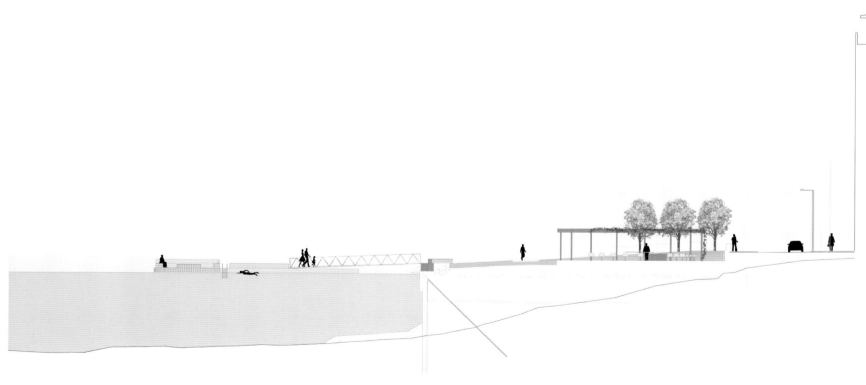

霍恩博格海滨公园荣获了 2012 年度瑞典景观奖锡耶纳奖。

霍恩博格海滨公园是一处美丽的所在 水陆相交在弯曲的河岸，景观设计现代感十足，拥有自然的圆形造型，线条干净利落。该公园向西面对着乌勒森达西安，傍晚时分还可见夕阳西下的美景。滨水岸区和三个长长的浮动码头让游客仿佛置身于水光交融之中。特别是炎炎夏日的午后，这里更成为了一片绿洲：周围的居民都来到这里晒太阳、游泳。公园里还有几处富有特色的休闲座椅区和一处淋浴场。淋浴场内有一个安置在高处的水槽，通过太阳能给水加热，可供慢跑者使用。

整个公园绵延超过 700 米，由四个部分组成。公园的西部有一个用作日光浴场的防波堤，木质船坞高高低低地插入湖中。东部的卡雅帕区与设计自然的河滨公园形成了对比。该区域呈水平圆盘状，轻微地向水面倾斜并略高于周围景观。更东边的部分是一个经过重建的原有景观，改造后的景观使游客更容易到达。

该工程还包括莫阿·马丁森广场。基于广场设计的建议，Nyréns 着重于两点：一是设计广场与乌勒森达西安交界处一个小景点的空间环境，二是为了突出作家莫阿·马丁森所进行的艺术装饰。为了构成通往大楼的通道，广场的地表高于周围路面。这种高度差形成了一面墙和几处台阶，可供游人休息。面向广场开放的楼梯，沿对角线向前延伸便到了湖边，而其前方便是卡雅帕区。

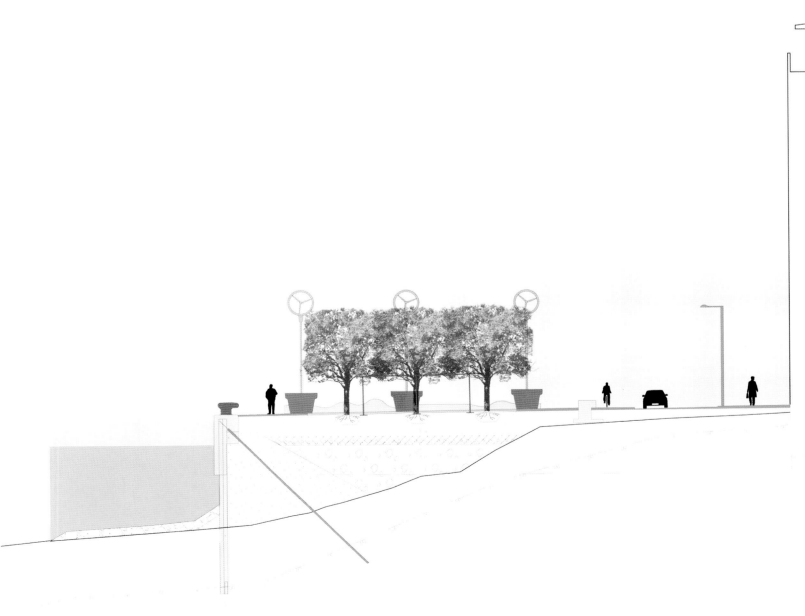

ARCHITECT Aedas   **TEAM** Tony Ang (Managing Director) / Gruffudd Owain (Project Director) / Jared Lee

SENTOSA BOARDWALK
SENTOSA GATEWAY, SINGAPORE

LANDSCAPE ARCHITECT Tierra Design (S) **CLIENT** Sentosa Development Corporation (SDC) / Singapore

The Sentosa Boardwalk is built on a new structure, over the sea and abutting the existing Sentosa Bridge, with an overall length of 670 metres. The Boardwalk mainly serves as a link to facilitate pedestrian flow between the two neighboring development, connecting the Sentosa Island with Vivocity seamlessly. Situated near the apex of the Southern Ridges, the Boardwalk design seeks to extend this greenbelt into the Sentosa Island in celebration of tropical living in Singapore. Five landscape themes which best represents the tropical mood - the mangrove swamp, rock garden, rolling hills and terrain, costal floral, and tropical rainforest - are introduced along the boardwalk to orchestrate a diversified environment to evoke an ever-changing experience along the bridge.

With an area of 16,400sqm, the timber deck is designed to push the boundaries of a bridge beyond a means of transport, by assuming a role as a public space, allowing users to rest, play and dine along a lush garden set on the sea front. To incorporate these spaces while maintaining the imagery of a timber bridge, the timber surfaces peel and fold organically to carve out spaces, seating and undulating terrain. This creates an aesthetic interplay of contrast between the intended rustic experiences, with the contemporary urban setting.

Situated along the sea front, the permeability and openness of the circulatory spaces ensures that users are constantly exposed to natural elements such as sun, sea breeze and lush greenery while visually connected to the sea view in the horizon. This seamless connection with adjacent development seeks to encourage users to step into the open and embrace nature. Night lighting created by light sculptures placed along the boardwalk is subtle at the start of the journey from Vivocity and progressively illuminates brighter towards the climax at the end of the Boardwalk. This creates a dynamic identity of the Boardwalk and a sense of anticipation of entering the vibrant Sentosa Island.

LANDSCAPE SCHEME

RESORT WORLD SENTOSA

ZONE E RAIN FOREST — Tall and lean rain forest trees, with the characteristics of different layers of green in the tropical, create very cooling and green spaces on the ground. Tall trees are used in this zone to minimize the blockage of view. The grand entrance archway to Resort World, as well as to provide shading to the plaza during daytime.

ZONE D COASTAL FLORA — Coastal flora is suitable to be grown in the marine area, and it requires little gardening job. These type of plants are usually at ground level, approx. 700mm in height. The flowers from these plants bring different fragrance to the surrounding.

A type of Lagerstroemia. This tree is commonly seen along the driveway in Singapore.

Revert-L shape of outdoor lighting, not only provides lighting in the dark, also creates dynamic perspective on the boardwalk. It gives a sense of direction to the public as they walk along.

Lagerstroemia trees have huge canopies, bright coloured flowers, and is commonly found in regional gardens and parks.

ZONE C TERRAIN & HILL — The terrain & hill concept is derived from this undulating landscape.

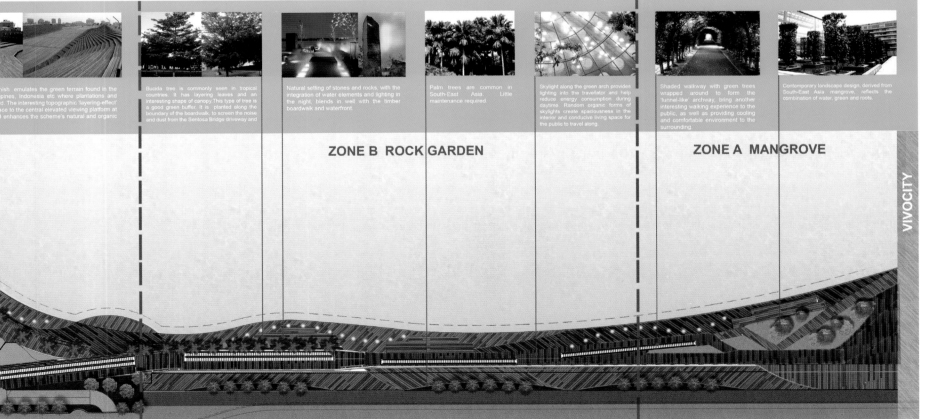

ZONE B ROCK GARDEN

ZONE A MANGROVE

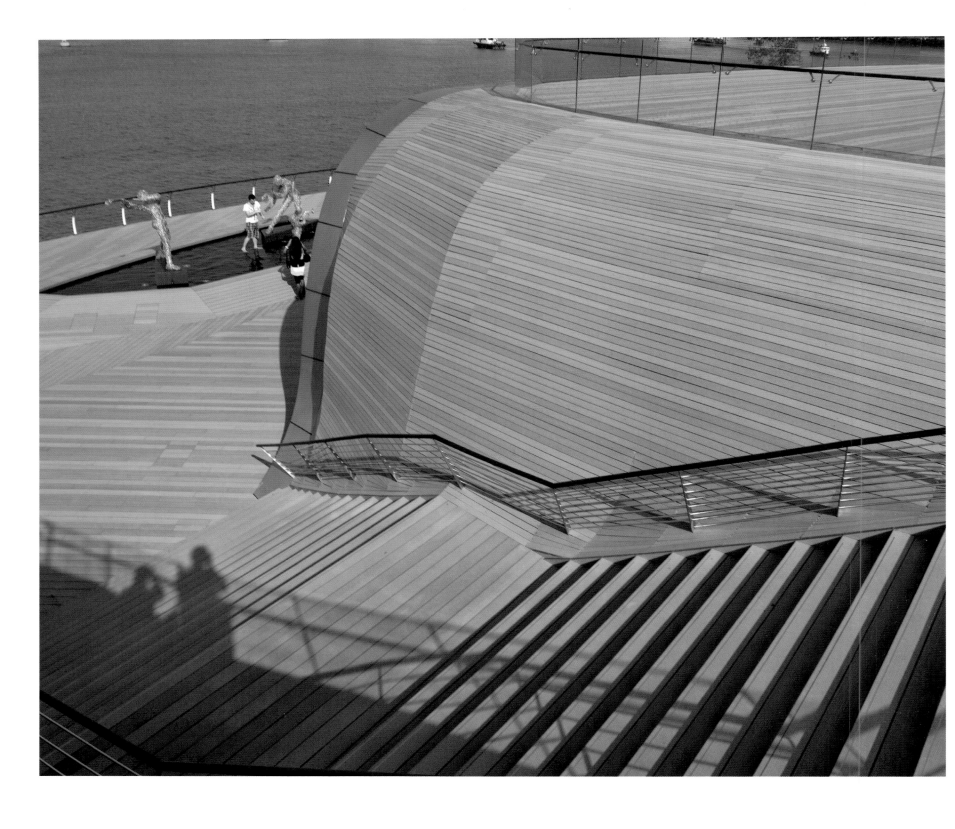
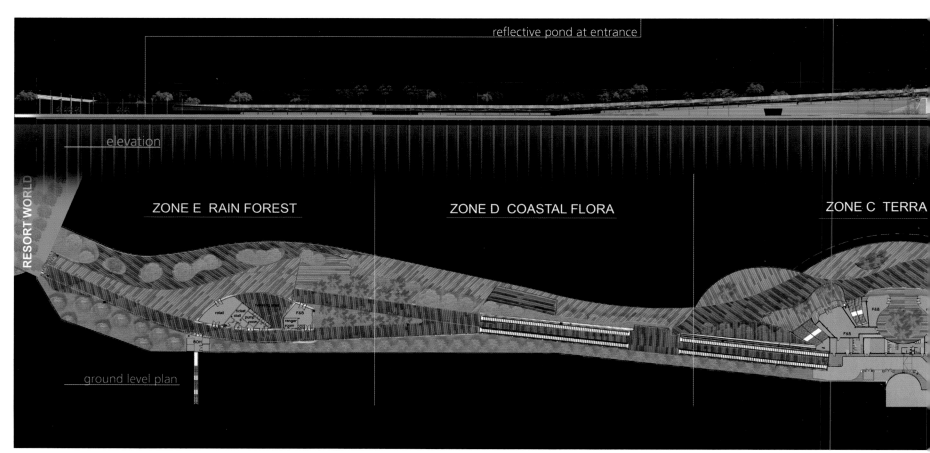
reflective pond at entrance

elevation

RESORT WORLD | ZONE E RAIN FOREST | ZONE D COASTAL FLORA | ZONE C TERRA

ground level plan

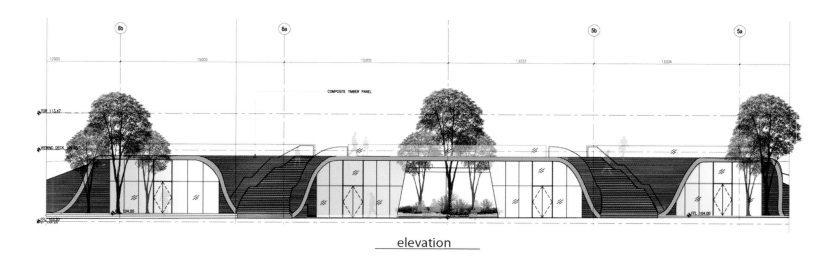

elevation

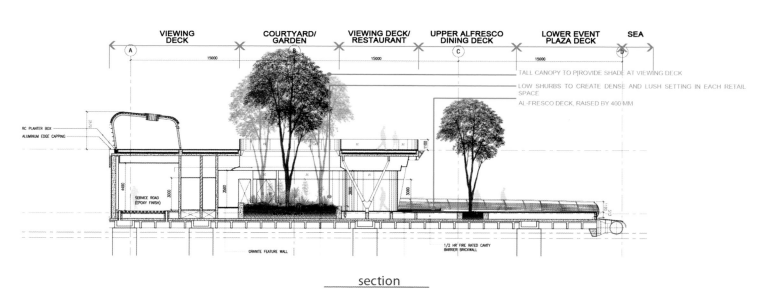

section

圣淘沙跨海步行道总长度为670米，毗邻原有的圣淘沙大桥，是新建的跨海通道。步行道实现了圣淘沙岛和怡丰城商场之间的无缝连接，为往来的行人提供便利。步行道位于南部山脊最高端，其设计力图延伸这片通往圣淘沙岛的绿色地带，以体现新加坡的热带风情。沿路有五个花园主题区：红树林湿地、岩石庭院、起伏的小丘陵、沿海植物区和热带雨林。这样的精心设计能够使人体验到多元化的环境，体验到步行道周边千变万化的景观。

占地16400平方米的观景木台的设计，使木板桥不仅仅局限为交通通道，而且可以作为一个公共的空间，为游客提供休息、娱乐和进餐的场所，同时，游客们还可享受海滨区的繁茂花园。为了在保证空间功能的同时充分展现木板桥的景致，木料表层经过自然干燥剥落，做成各式的空间、座椅和波动起伏的地势。如此设计出的乡村质朴风与现代城市环境之间形成对比，具有审美冲击。

滨海区的地理位置、循环空间的渗透性和开放性确保了游客既能看到海平面上的风景，又可以接触到诸如阳光、海风和茂盛的绿色植物等自然元素。这种与相邻地区天衣无缝的结合旨在鼓励游客接近和拥抱大自然。沿着步行道而建的灯光雕塑提供了夜间照明，从怡丰城商场开始，灯光渐渐由弱变强，到步行道的尽头灯光达到最强。这样的灯光效果使步行道产生了动态的变化，也让人充满期望地走进活力四射的圣淘沙岛。

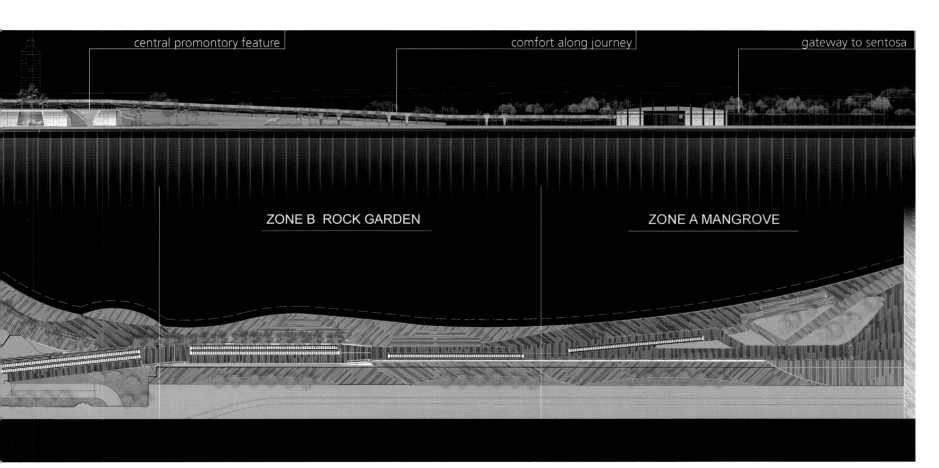

Overall Plan

Section across travellator surrounded by various strategies of green wall and planters

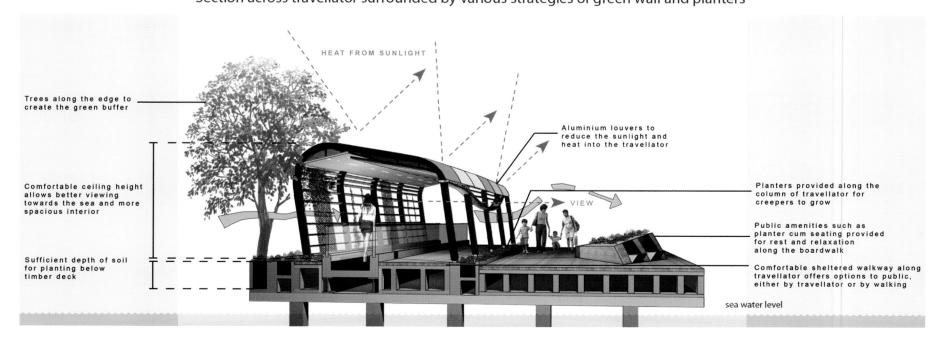

LANDSCAPE ARCHITECT In Situ   AREA 100,000 m²

LYON RIVER BANK
/ RHÔNE-ALPES, FRANCE

里昂河岸

The left bank of the Rhône is a 10 hectares site at the very heart of the city which is in direct contact of the river over a total distance of almost 3 miles.

Previously used for car parking, the project is a true melting pot for ideas concerning city-centre development. The regeneration of the left bank of the Rhône forms part of a wider project for social and urban renaissance.

The Berges du Rhône development seeks to give city-dwellers a variety of areas in which they can relax and feel closer to nature. It also plays a role in the city's urban transport plan, which aims to reduce car usage for daily journeys by encouraging the use of sustainable modes of transport. This, in turn, has allowed the regeneration of public spaces.

When the first sections were opened to the public in spring 2007, the development proved an immediate and massive success. Today it is a space that has already firmly taken root in the day-to-day life of Lyon's residents.

séquence 01: Le Bretillod et la Lône

séquence 03: La Rive Habitée et l'Archipel des Iles Jardins

séquence 04: La rive active

séquence 02: La Ripsylve Amont

tive, la Grande Prairie du Rhône et les Planches   séquence 05: Les Terrasses de la Guillotière

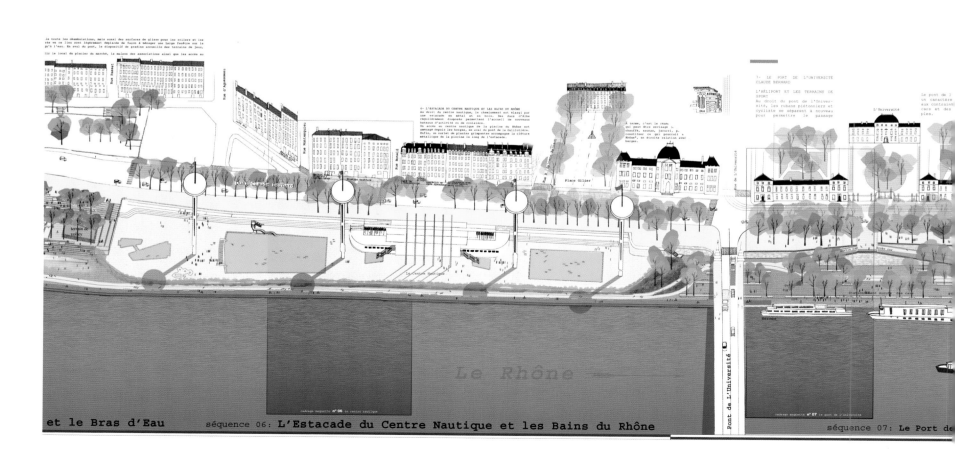

séquence 06: L'Estacade du Centre Nautique et les Bains du Rhône

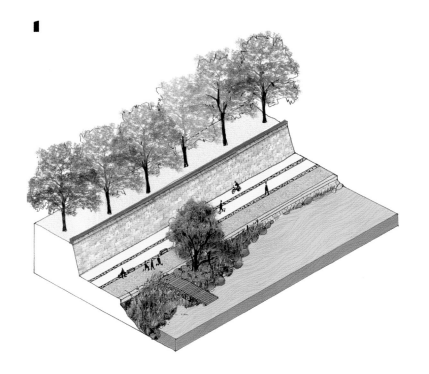

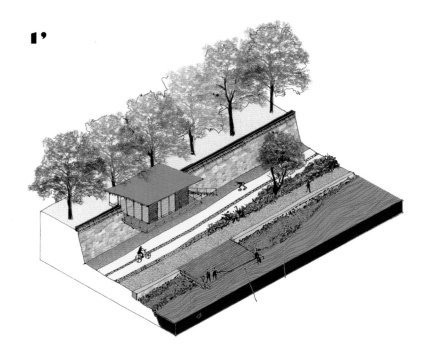

位于罗纳河左岸的这块 10 公顷的土地正处于里昂市中心，与罗纳河直接相邻，到河岸距离近 5 千米。

这块土地曾经是一个停车场，而现在是一个名符其实地汇集了各种城市中心开发理念的大熔炉。罗纳河左岸的重建是社会及城市复兴大工程的一部分。

罗纳河岸的开发为城市居民放松身心和亲近自然提供了多种空间，也为市内交通提供了便利。城市交通规划力图鼓励人们使用可持续的交通方式，在日常出行中减少汽车的使用。而反过来，这又促进了公共空间的复兴。

2007 年春季，第一批项目工程建成并对公众开放，立即获得了巨大的成功。如今，这里已经成为里昂居民日常生活中必不可少的一部分。

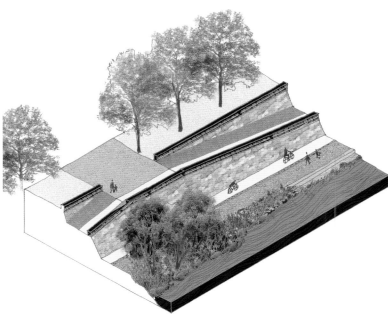

2'

2'

séquence 08 : Les Jardins du Rhône, la Gal

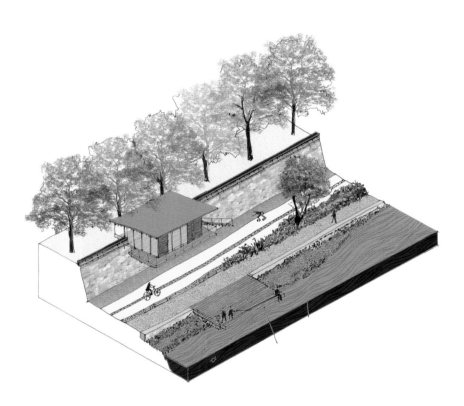
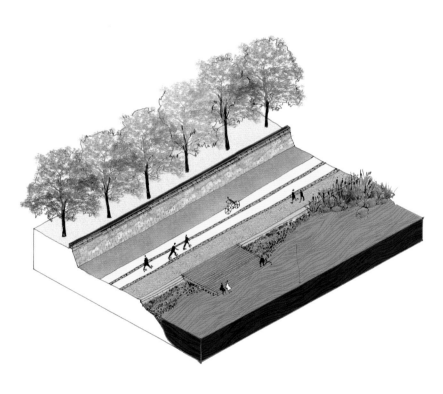

9- LA LIAISON VERS LE PARC DES BERGES ET GERLAND
La liaison avec le parc des berges et le parc de Gerland est étudiée en option jusqu'au pont Pasteur.
Cet aménagement consiste à prolonger les différents rubans de promenade en élargissant le cheminement actuel. La berge fait l'objet d'un reprofilage et d'une végétalisation plus adaptée et naturelle.

SENIOR LANDSCAPE ARCHITECT Edwin Santhagens   **LANDSCAPE DESIGNER** Stephen Tas / Wim van Krieken

SCHINKELEILANDEN
/ AMSTERDAM, THE NETHERLANDS

辛克尔小岛

The Schinkel zone is an area along the Schinkel channel, located on the fringe of Amsterdam.

The zone is interpreted as an archipelago existing of four different islands that house a large number of recreational facilities. The backbone of the design is a recreational and ecological connection between the Vondelpark and the Amsterdam Forest, using an old railway embankment. The various functions determine the nature of the islands: There is a tennis island, an island of boat houses, a soccer island, a park island and a nature island. By increasing the total area of water and using piers, platforms, bridges and reed beds, the presence of water is felt more directly.

The area becomes a unique Water Park forming a new entrance for the city of Amsterdam.

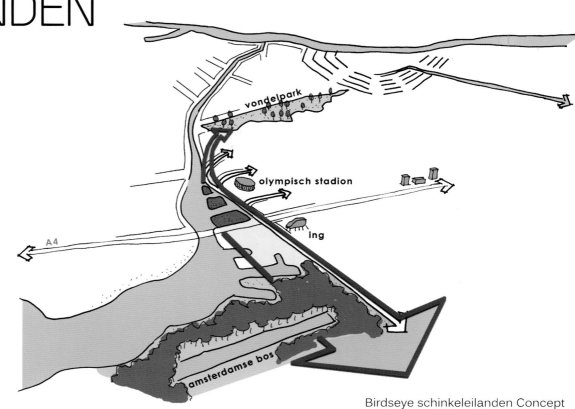

Birdseye schinkeleilanden Concept

FIRM Buro Sant en Co   **AREA** 170,000 m²   **AWARD** Winner of the (Dutch) Green City Award 2011

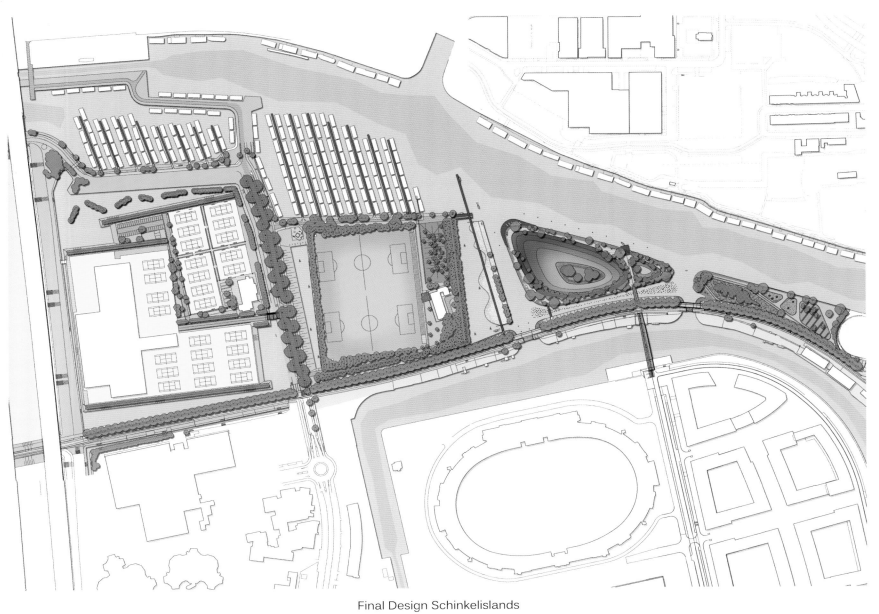

Final Design Schinkelislands

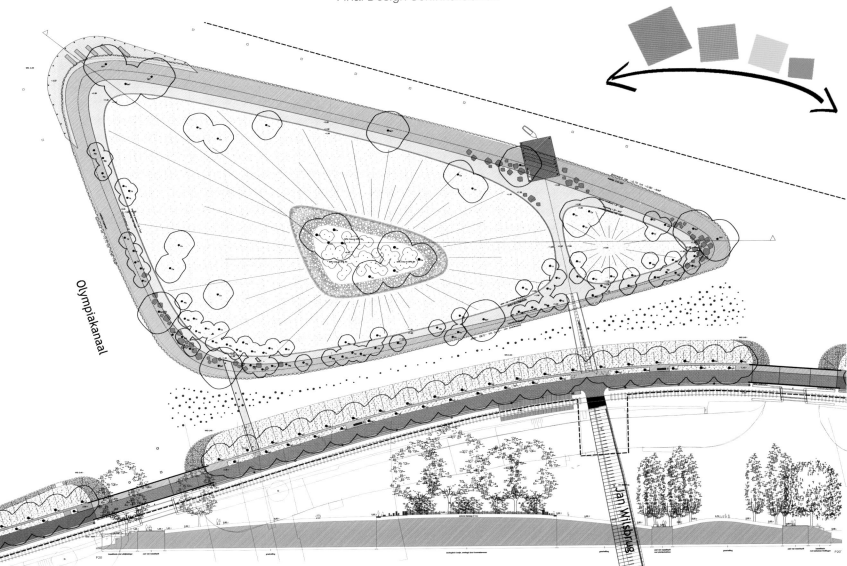

Final Design Park Island

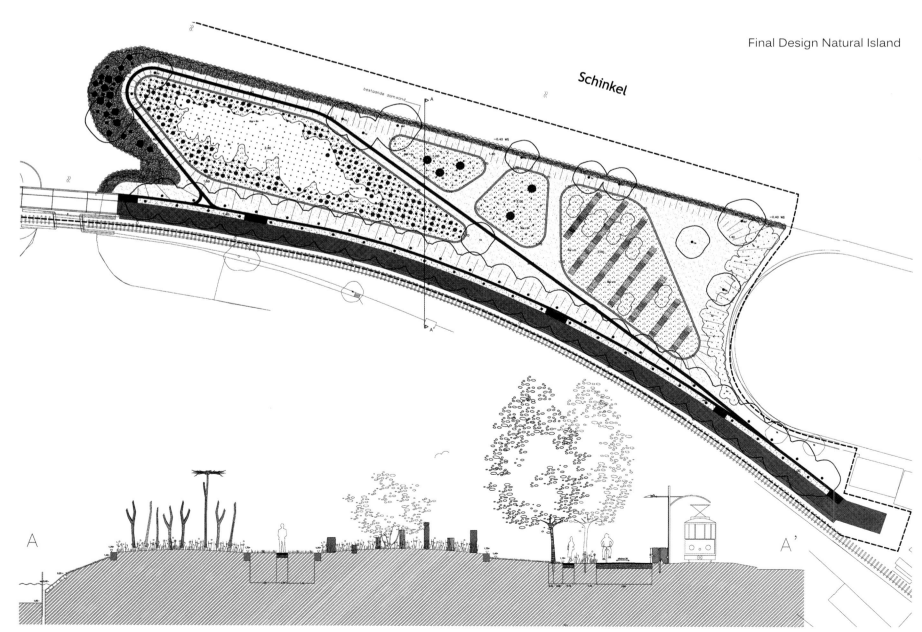

Final Design Natural Island

Schinkel

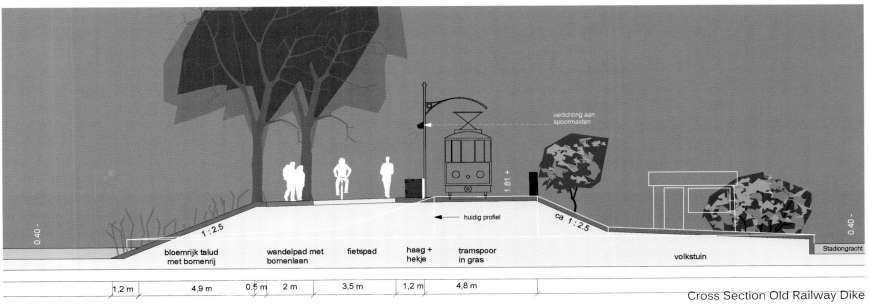

	bloemrijk talud met bomenrij	wandelpad met bomenlaan		fietspad	haag + hekje	tramspoor in gras	volkstuin	
1,2 m	4,9 m	2 m	0,5 m	3,5 m	1,2 m	4,8 m		

Cross Section Old Railway Dike

辛克尔地区位于辛克尔运河沿线，坐落于阿姆斯特丹边缘。

该地区是由现存的四个不同岛屿组成的群岛，岛上有许多休闲娱乐设施。设计中保留了旧的铁路路堤，其核心目的是实现冯德尔公园与阿姆斯特丹森林之间娱乐休闲和生态保护的结合。不同的功能决定了这些岛屿的特质：这里有可以打网球的岛屿，有船坞式的岛屿，有可以踢足球的岛屿，有公园式的岛屿，还有可以亲近大自然的岛屿。通过扩大水域面积和使用码头、平台、桥梁和苇丛河床，人们可以更近距离地接触到这片水域。

这片区域成为了一个独特的水上公园，形成了一个通往阿姆斯特丹的新入口。

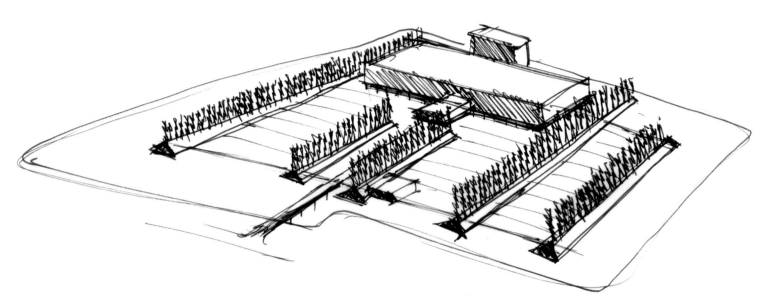

LANDSCAPE ARCHITECT/URBAN DESIGN Taylor Cullity Lethlean / Peter Elliott **CLIENT** Lend Lease Development

VICTORIA HARBOUR, DOCKLANDS

/ MELBOURNE

达克兰的维多利亚港

PHOTOGRAPHER Rhiannon Slatter / Emily Taylor

Taylor Cullity Lethlean and Peter Elliott were engaged to prepare a master plan and subsequent conceptual development and implementation of Victoria Harbour's principal public promenade and adjacent urban spaces. A comprehensive phase of analysis and modelling established a program of activities including marinas, cafes, water plazas and event spaces. These elements are structured within a design strategy that celebrates this site as one of "movement, event and destination". This intent is realised by establishing a sequence of sculpturally formed meeting places that enrich the experience of their waterside context.

Victoria Green is a small urban park in the heart of the Central Precinct. It is looked upon by the surrounding high-rise commercial and residential buildings, and aims to cater for the associated demographic.

The main components of the park are the sunken elliptical lawn, encircled by garden bed; the timber deck seating and picnic zone; and overriding ring of Eucalyptus trees. The main ramp leading off the deck is enframed by perforated steel screens whose shadow patterns spill onto the surrounding deck and grass throughout the day.

The design of Water Plaza, ANZ precinct and associated promenades included extensive reshaping of the former shipping Wharf with sculptured lawn planters, sweeping timber decking curving around original timber piles within the water, seating pods and integrated artworks by Mark Stoner.

The reworking of a "green overlay" into the master plan meant the extensive paved areas of the upper promenade are complemented by broad expanses of grass, an avenue of mature Peppercorn trees and distinctive vegetation in planter boxes along the edge of a cafe zone. Water sensitive urban design principles are employed and the harvesting of stormwater is used in the sites irrigation system.

The aim of the project is to create an urban waterside environment as the centrepiece of the public realm in the Victoria Harbour Precinct.

泰勒·考利特·莱斯利恩和彼得·艾略特对维多利亚港的主要公共步行区和毗邻的城市空间进行了工程的总体规划设计、后续概念的开发和实施。经过综合分析和建模，设计师们提出了一套建设方案，其中包括多个游船泊位、咖啡厅、水上广场和休闲活动空间。这些都迎合了将维多利亚港打造成"活动中心、庆典场所和旅行目的地"的设计构想。经过缜密的设计，先后建成了几座会议中心来实现这一构想，同时，滨海风景区也因此而丰富多彩。

维多利亚绿地中心位于维多利亚港辖区腹地，是一个小型城中公园。附近的商用和住宅高楼将公园围在中间，那里的人们可以来此休闲娱乐。

公园的主要组成部分包括：一个由花坛环绕的、中心下沉的椭圆形草坪，木栈道座椅和野餐区以及重要的环形桉树林。木栈道通向被钢板围住的主坡道，钢板上有许多孔洞，阳光透过这些孔洞投射在坡道周围的木栈道和草坪上，一整天都不间断。

在设计水上广场、ANZ办公辖区及附近的步行区时，设计师将之前的码头进行了大规模的改造，以景观草坪代之。他们还沿着水中木桩的位置，建起了水边蜿蜒的散步路和休息座椅，还陈设了与环境浑然一体的由马克·斯通纳创作的艺术品。

在修改总体规划时，"整体绿色覆盖"的概念被加入其中。这就意味着大面积方砖铺路的上游步道与宽广的草坪相映成趣。沿着咖啡厅的道路上栽种着成熟的胡椒树，方盒花坛中长着奇花异草。节水的概念也在城市设计有所体现，整个景区的灌溉均使用暴雨时收集的雨水。

这一工程的目的是在维多利亚港辖区的公共用地上创造一个城市滨水景观，使其成为该区域的重点景观。

LANDSCAPE ARCHITECT Kevin Shanley **FIRM** SWA Group
CLIENT / OWNER Shenzhen Urban Planning and Land Administration Bureau / Shenzhen Public Works Bureau / Shenzhen Green Administration Bureau

SHENZHEN BAY COASTLINE PARK
/ SHENZHEN, CHINA
深圳湾海岸公园

AREA 1,030,000 m²   **PHOTOGRAPHER** Tom Fox and Jonnu Singleton of the SWA Group

The design and construction of the Shenzhen Bay Waterfront established a permanent, public waterfront park for 10 million people, gives new life to waterfront as an irreplaceable ecosystem, reclaims a South China cultural landscape, and serves as a model for sustainable waterfront development.

Shenzhen, China is located in the Pearl River Delta in a region known for its complex network of rivers that meet at a mountain fringed sea. Historically a coastline of small fishing villages, in 1980 the region was established as a Special Chinese Economic Zone and it entered into a thirty year period defined by uncompromising "modernization". Huge monetary wealth was brought to the region by mining entire mountains in order to fill its oceanside bay and to rapidly create more developable land. Generations of people who defined their sense of place by a bay connection were left to wonder how the water had so quickly vanished from beneath them.

Shenzhen has become an internationally recognized city through bold ambitions, much persistence, and a singular focus on creating hard edges of modernity. However, the price of this approach reached a crescendo louder than the cranes, dump trucks, and construction crews enveloping the city. An international design competition was held and SWA's entry was selected based on their concept of creating a final landfill line focused on "edges, connections, and cycles" honoring the bay's historic sense of place. SWA's landscape architecture team subsequently developed the project from planning concepts to design development as well as providing construction phase services.

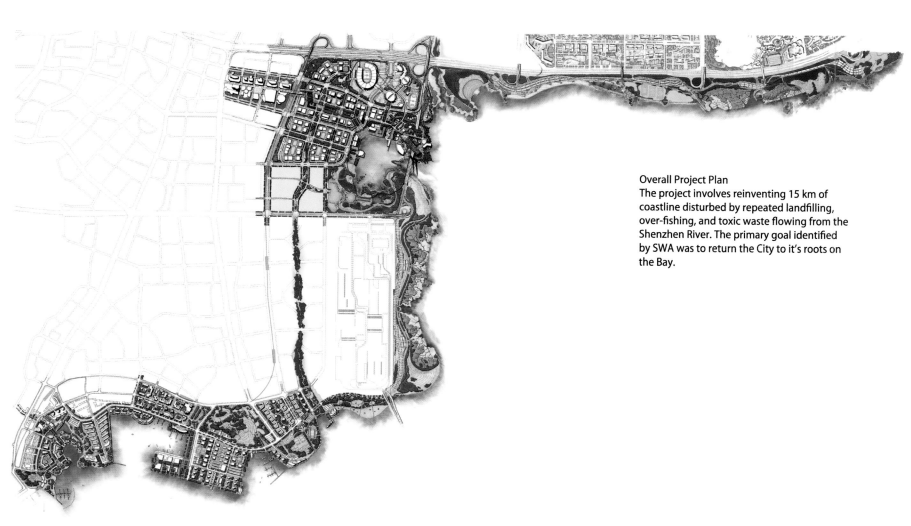

Overall Project Plan
The project involves reinventing 15 km of coastline disturbed by repeated landfilling, over-fishing, and toxic waste flowing from the Shenzhen River. The primary goal identified by SWA was to return the City to it's roots on the Bay.

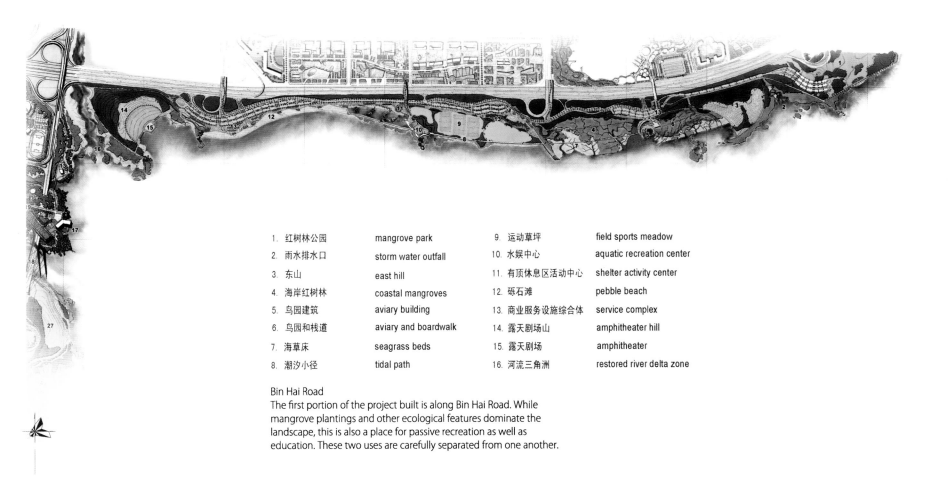

1. 红树林公园	mangrove park	9. 运动草坪	field sports meadow
2. 雨水排水口	storm water outfall	10. 水娱中心	aquatic recreation center
3. 东山	east hill	11. 有顶休息区活动中心	shelter activity center
4. 海岸红树林	coastal mangroves	12. 砾石滩	pebble beach
5. 鸟园建筑	aviary building	13. 商业服务设施综合体	service complex
6. 鸟园和栈道	aviary and boardwalk	14. 露天剧场山	amphitheater hill
7. 海草床	seagrass beds	15. 露天剧场	amphitheater
8. 潮汐小径	tidal path	16. 河流三角洲	restored river delta zone

Bin Hai Road
The first portion of the project built is along Bin Hai Road. While mangrove plantings and other ecological features dominate the landscape, this is also a place for passive recreation as well as education. These two uses are carefully separated from one another.

Ecological Education Center
The eastern end of the Bin Hai section includes an ecological education center the focuses on the restoration of Shenzhen Bay that is being done in both Hong Kong and Shenzhen. A large viewing hill protects the mangrove and seagrass beds while interpreting the landforms that were once part of the Shenzhen coastline.

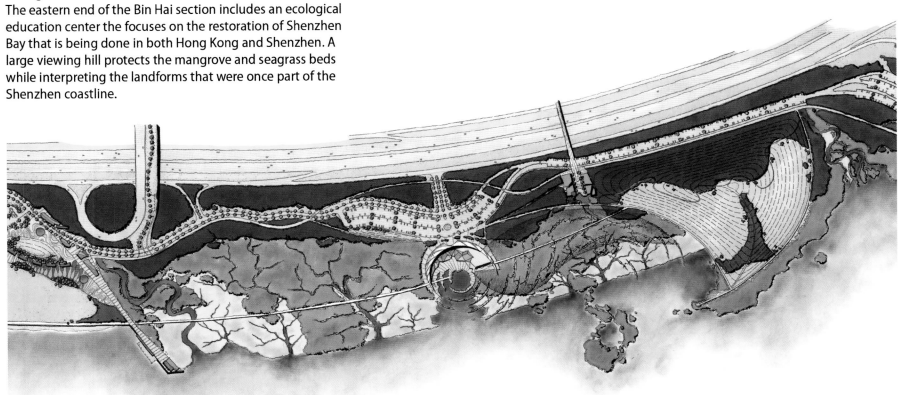

Beach and Aquatic Center
A long sandy beach once framed views of Shenzhen Bay. This section of Bin Hai seeks to rebuild a portion of that forgotten beach. A saltwater swimming pool and aquatic center anchor an active sports zone.

Nanhi Development Zone
The western most end of Bin Hai Road is anchored by a mixed use development and entertainment zone. This zone is informed by a large tidal basin and a series of stormwater canals, basins and other features that bring natural systems into the fabric of the project.

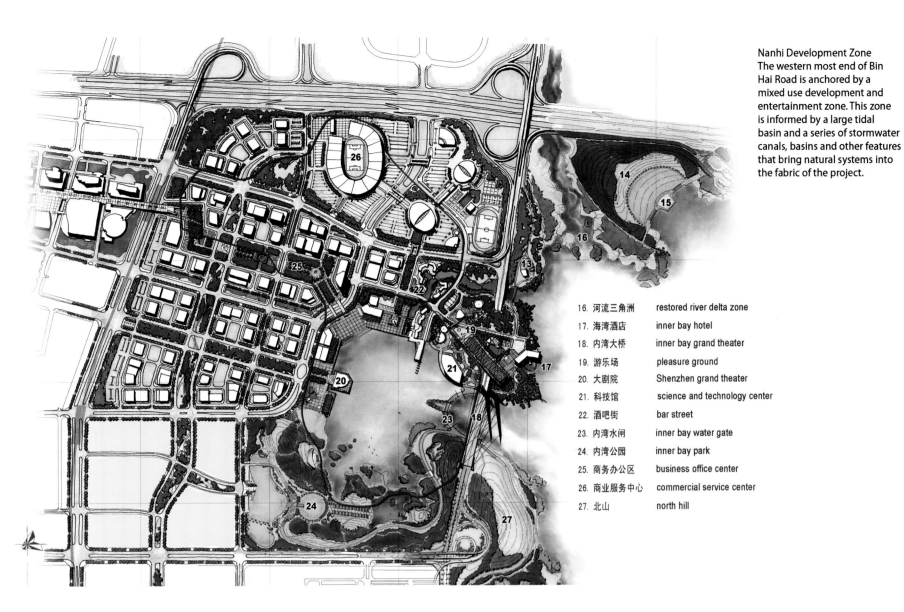

16.	河流三角洲	restored river delta zone
17.	海湾酒店	inner bay hotel
18.	内湾大桥	inner bay grand theater
19.	游乐场	pleasure ground
20.	大剧院	Shenzhen grand theater
21.	科技馆	science and technology center
22.	酒吧街	bar street
23.	内湾水闸	inner bay water gate
24.	内湾公园	inner bay park
25.	商务办公区	business office center
26.	商业服务中心	commercial service center
27.	北山	north hill

Sheko Zone
The most southern portion of the project lies within an urbanized area. The design of this zone seeks to carry the grid of the city down to the waters edge. A new waterfront "corniche" is created to provide new economic activities while giving the city a series of windows on bay.

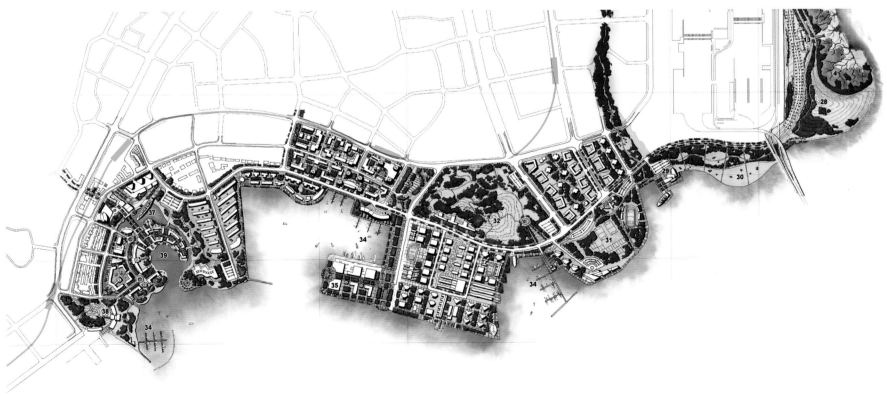

28.	大桥山	bridge hill	34.	游艇码头	boat dock
29.	海事博物馆	maritime affairs museum	35.	渔人码头	fish pier
30.	大桥森林公园	bridge hill park	36.	蛇口老镇商业中心	shekou traditional down town
31.	体育公园	sports park	37.	明华轮	minghua ship
32.	蛇口山	shekou hill	38.	南海酒店	nanhai hotel
33.	蛇口广场	shekou plaza	39.	海上世界	ocean world

/ 149

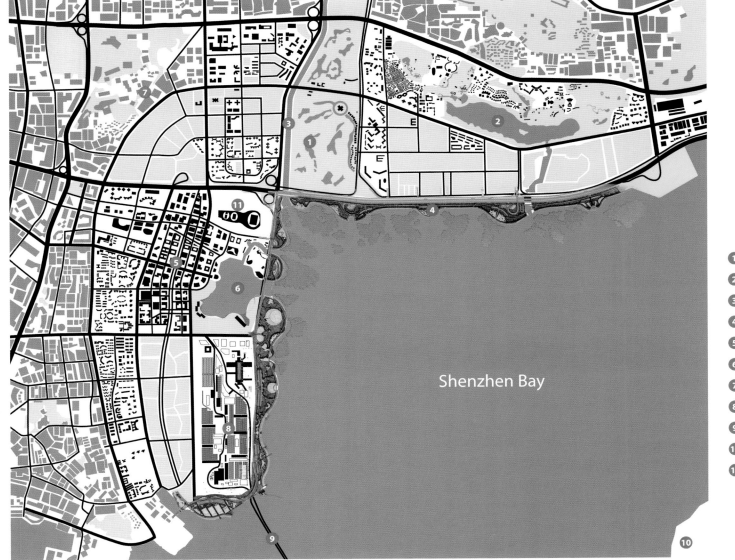

1. Shahe Golf Court
2. Oversea Chinese Town
3. Dashahe River
4. Binhai Boulevard
5. Nanshan Houhai CBD
6. Nanshan Inner lake (F1 boat racing)
7. Shenzhen University
8. Land Port
9. Hong Kong Bridge
10. Hong Kong
11. Nanshan Sport City

Shenzhen Bay

　　深圳湾海滨区的设计与施工给 1000 万市民提供了一个永久性的公共海滨公园。作为不可替代的生态系统，该设计也给海滨区带来了新的活力，成为中国南方的一个人文景观，也是水域可持续发展的典范。

　　深圳位于中国珠江三角洲地区，该地以若干条江河入海形成的复杂地势而闻名。历史上，深圳曾经是一个小渔村，在 1980 年被确立为中国经济特区。30 年来，一直坚持走现代化道路。当地全面开采山脉用来填充海湾，快速扩大开发土地，所以集聚了巨额资金。由于海湾的存在，这里的人们世世代代都靠它寻找方向。可现在的人们只能慨叹，脚下的水为何消失得如此之快。

　　深圳抱负远大、坚持不懈，高度重视创造现代化前沿技术，已经发展成为国际公认的大都市。然而，深圳付出的代价就是成为了日渐喧嚣的城市：遍地都是起重机、翻斗卡车和建筑施工队。深圳举办过一次国际设计大赛，SWA 从中脱颖而出，以"海岸、连接和循环"为关键点，以建设最终堆填海岸线为理念，从而把海湾建成富有历史感的景观。随后，SWA 景观设计咨询有限公司团队从规划理念到设计开发，做出诸多规划，并在施工阶段提供了多项服务。

LANDSCAPE ARCHITECT Antonio Lorén / Eduardo Aragüés / Raimundo Bambó   **FIRM** ACXT Architects

THE INTERVENTION IN THE EBRO BANKS

/ TENERÍAS PARK, LAS FUENTES, ZARAGOZA, SPAIN

埃布罗河畔的干预项目

AREA 87,535 m²   **PHOTOGRAPHER** Aitor Ortiz

The north boundary of the Tenerías - Las Fuentes project area covered the space along the right bank between the Ebro river bed and the buildings that comprise the north edge of the Tenerías and Fuentes districts, running as far as the location of the future dam (calle Fray Luis Urbano) to the east and the Puente de Hierro (Iron Bridge, from the XIX century) to the west.

This is a built-up area consisting of the Echegaray and Caballero promenade and a park running lengthwise between this avenue and the river Ebro.

The possibility of creating a walkway along the right bank of the Ebro through a garden area running from the iron bridge along Echegaray promenade to near the bridge on the third ring road, taking into account the future Ebro dam pedestrian link, makes this project an outstanding route for pedestrians through the city.

The project creates a pedestrian link between the centre of the city and the natural setting of Soto de Cantalobos and the river Gállego water park, and creates a network of routes running the length of the project area, clearly differentiating between the roads and paths for cars, bicycles and pedestrians in the city, and those for water, sports, leisure and pedestrian activities in natural settings.

The project enhances the view from this area towards the Virgen del Pilar's Basilica, creating a link between the project area and the city centre thanks to one of its most famous landmarks.

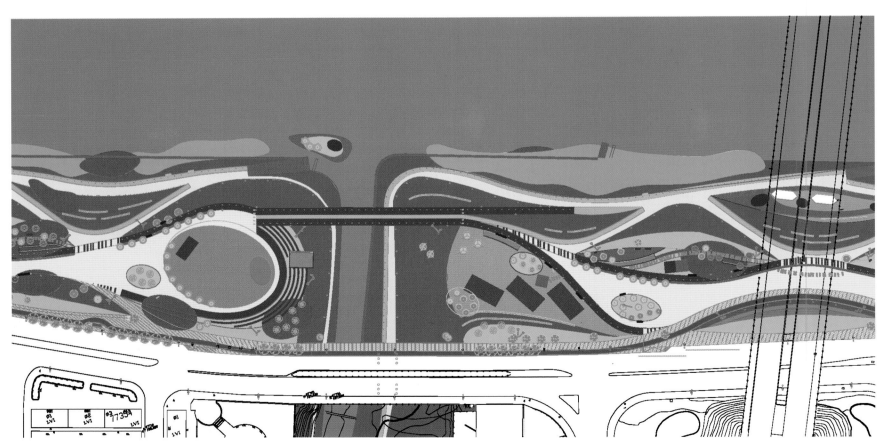

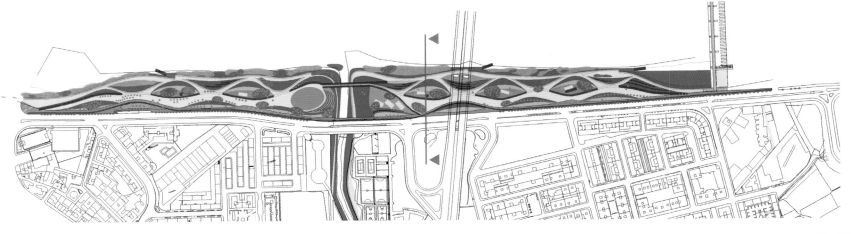

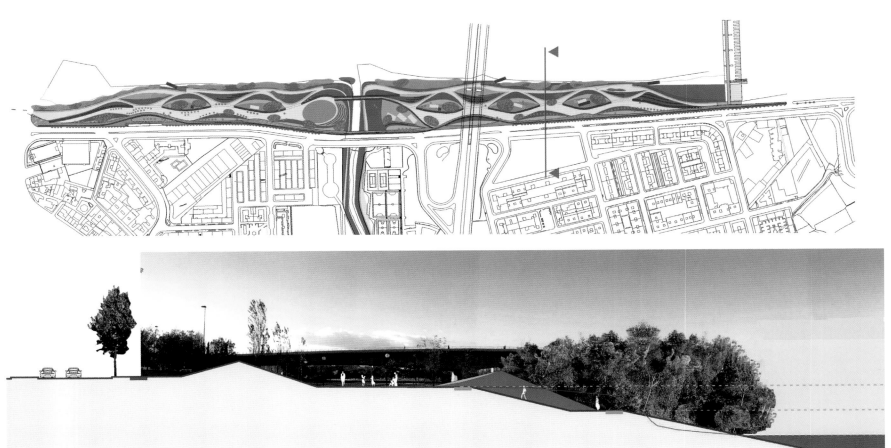

/ 163

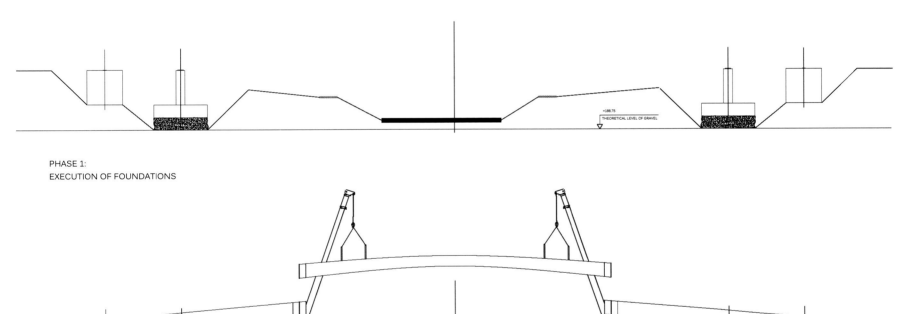

PHASE 1:
EXECUTION OF FOUNDATIONS

PHASE 2:
PLACING OF COMPENSATION OPENINGS
ANCHORAGE TO FOUNDATIONS
EXCAVATION INFILL
PLACING OF CENTRAL BEAM

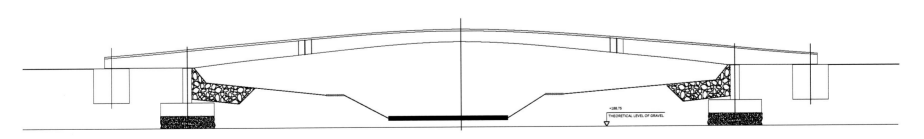

PHASE 3:
WELDING OF CONTINUOUS BEAM
SETTING UP OF SHUTTERING SLABS
CONCRETE POURING OF UPPER SLAB
PLACING OF PROTECTIVE ROCKFILL

MIRROR PLAN VIEW

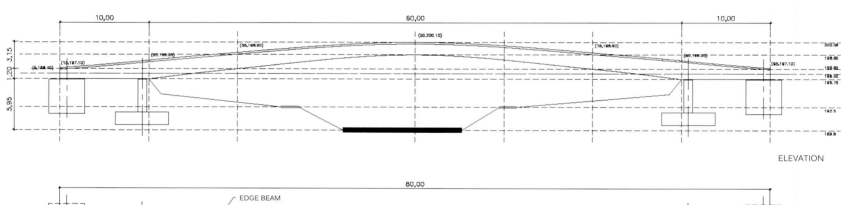

ELEVATION

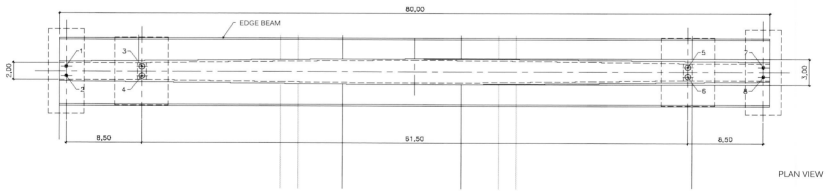

PLAN VIEW

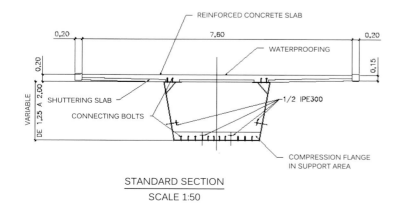

STANDARD SECTION
SCALE 1:50

该项目 Tenerías 至 Las Fuentes 段北起埃布罗河河床的右岸及建筑群之间，向东可以延伸到规划中的水坝（弗赖路易斯乌尔巴诺水坝），西至 Puente de Hierro 桥（19世纪铁桥）。

这一区域由埃切加赖和卡瓦列罗步行道以及一个在步行道与埃布罗河之间纵向延伸的公园组成。

该项目造就了一条穿城而过的绝佳步行线路，它沿着埃布罗河右岸，穿过花园区，以铁桥为起点，顺着埃切加赖步行道一直延伸到位于第三环形路上的大桥附近，将来还可以与埃布罗大坝步行路相连。

该项目在市中心和 Soto de Cantalobos 自然景观与 Gállego 水上公园之间建造了一条人行通道，区域内道路四通八达，清晰地分为汽车路、自行车路和人行路，自然景观中也划定了水域、运动、休闲及散步活动区域。

从项目向皮拉尔圣母大教堂眺望，风景秀美，项目将本区域与市中心联系起来，使它成为了该地区最有名的地标之一。

LANDSCAPE ARCHITECT scape Landschaftsarchitekten GmbH   **CLIENT** Stadt Wetter – Stadtplanungsamt

SEEPLATZ WETTER
/ GERMANY

威特市斯博拉茨

The "Seeplatz" on Harkort Lakeside is the central public space of the city "Wetter (Ruhr)". At this place the town "Wetter" presents itself to the residents and guests of the Ruhr valley as a loveable city on the river. For the special situation of the square between the city and the lake a multifunctional urban stage was developed. The wooden deck is the terrace and the figurehead of the city. Here the city "Wetter" clearly opens itself to the lake, events can be celebrated, people are invited to go for a walk or rest on a bench watching the coming and going.

The square is paved throughout at ground level. The surface was fitted to the existing topography. A uniform slope to the bank emphasizes the open water area of the lake. The riverside walks are fast asphalt paths that are most suitable for inline skating, cycling or walking. The central route between the city and lake is transformed into a pedestrian promenade.

The lakeside terrace is the main target, viewing and lounge area of the square. Parallel to the water's edge, there is huge wood deck. In its center are two seating objects. The bench elements are developed in a soft momentum in the form of a bulge of the wooden deck. Their supporting elements are also used to integrate the lighting effects. The existing trees form the basis of the planning. Due to selective replanting or felling of trees, the intended spatial impact will be significantly strengthened.

AREA 6,750 m² **PHOTOGRAPHER** Matthias Funk

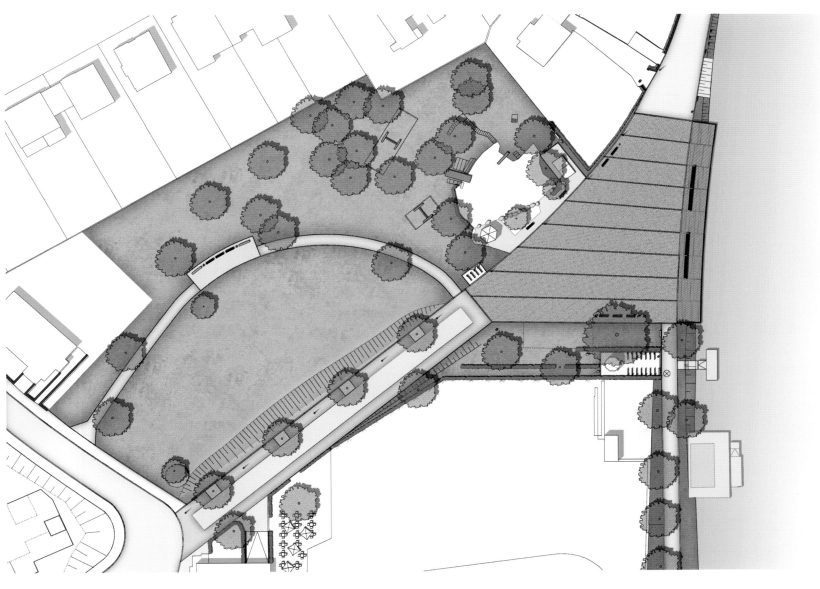

哈考特湖滨景观上的"斯博拉茨"是鲁尔地区威特市的中央公共区。鲁尔河谷的当地居民和游客都可以在这里欣赏到鲁尔镇这座惹人喜爱的河上之城的旖旎风光。由于"斯博拉茨"广场的特殊位置——位于城市与湖泊的中间地带，这里被开发成一个多功能城市空间。湖边的木栈道状若梯田，如果把鲁尔城比作一条大船，那么这些木栈道就如同船头的艏饰像。在这里，鲁尔城与湖水紧密相连，人们可以在此举行盛事，也可以漫步湖边，或是择席而坐，看着来来往往的旅行者。

广场的地面被铺成与地面一平的高度。地面与当地的地形相一致。一个向河岸均匀倾斜的斜坡凸显了湖泊中的开放水域。河边的漫步道是一条条便捷的沥青小路，这样的小路最适合轮滑、骑自行车或散步。位于城市与湖泊之间的中央通道则被改造成了步行长廊。

湖边的阶地则是游人主要的去处，人们可以在那里欣赏风景，也可将那里当作广场的休息区。与湖边平行的是巨大的木栈道。在木栈道的中央，有两处坐席。设计师采用一种柔和过渡的方式将长椅设计成木栈道上自然的凸起。支撑长椅的元素也被运用到设计当中，以便与灯光效果融合。景区整个的规划以原有的树木为基础。由于对树木进行了选择性的重植和砍伐，大大增强了人为营造的空间效果。

LANDSCAPE ARCHITECT Margot Long   **FIRM** PWL Partnership Landscape Architects Inc

SOUTHEAST FALSE CREEK
/ VANCOUVER, BC

东南福溪可持续社区

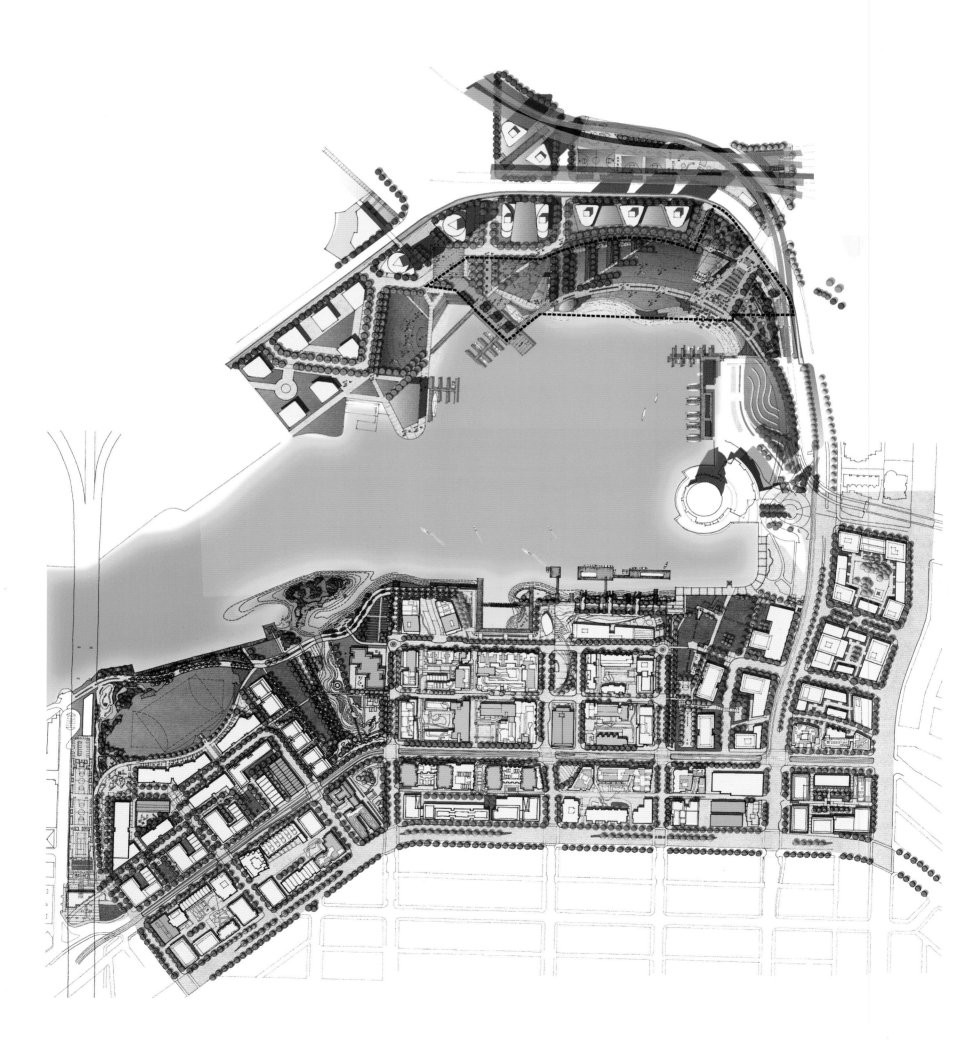

AREA 320,000m² **PHOTOGRAPHER** PWL Partnership Landscape Architects Inc

The Southeast False Creek (SEFC) Neighborhood is one of Canada's leading sustainable communities and Vancouver's first comprehensive sustainable neighborhood development. At approximately 32 hectares it will provide a mix of land uses, be home to 10,000 – 12,000 people in market and non-market housing, and demonstrate exemplary practices in energy and water conservation, innovative infrastructure practices and transit oriented development.

Waterfront Park - Phase 1 represents the first phase of SEFC's primary park and open space system and a 650-metre extension of Vancouver's iconic seawall. Through walkways, bicycle paths, diverse seating and gathering areas the park provides a variety of vital green spaces that will reconnect people with the heritage-rich waterfront and offer unique and memorable experiences unlike any other in Vancouver.

The waterfront design aesthetic reflects the intersection between the site's industrial legacy and the desire for a sustainable, contemporary urban park. An array of imaginative, universally accessible social spaces combine with various design elements to honour the site's shipbuilding heritage, including a "canoe-shaped" pedestrian bridge crossing a small inlet. Along with the stormwater management and plantings seen throughout the park, the inlet features sentinel lights that line the inlet and is fitted with special panels to prevent light pollution, and custom litter receptacles have solar-powered compacters to reduce the number of vehicle trips required to service them.

Waterfront Park – Phase 1 completes the existing waterfront promenade and bike lanes found throughout the False Creek and Downtown Vancouver areas. This innovative project demonstrates that environmentally sensitive design is completely at home in a dense residential and commercial neighborhood and shines even brighter when layered with historical references, social spaces, and recreational opportunities.

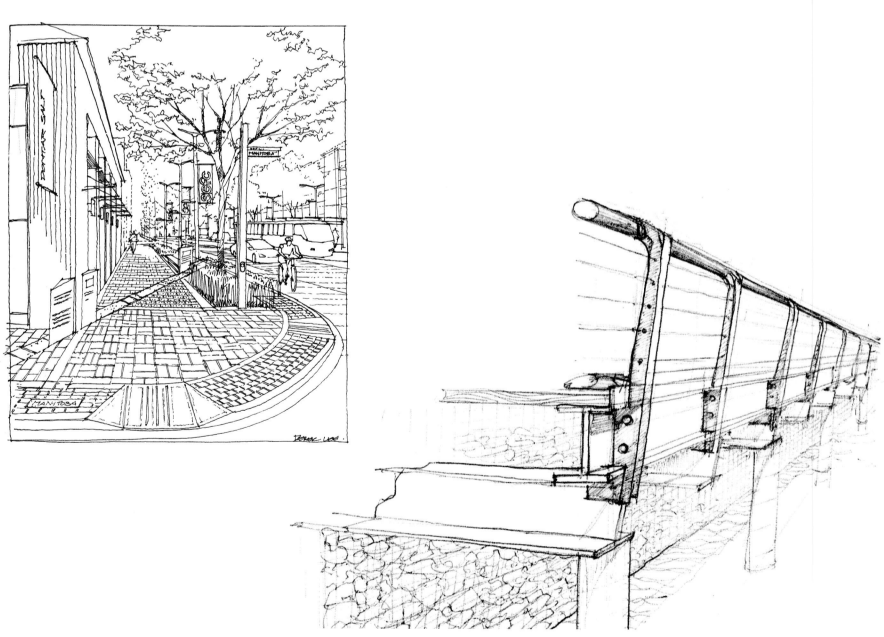

　　东南福溪（SEFC）住宅区是加拿大首屈一指的可持续社区，同时也是温哥华首个综合性的可持续社区开发项目。项目竣工后，面积约32公顷的综合用地将涵盖一个可容纳10000人至12000人的商业和非商业居民区，使其成为能源和水资源保护、基础设施创新和改造型开发项目的最佳典范。

　　水滨公园一期是东南福溪公园和绿地系统工程中的一期工程。同时，在该工程阶段将建成一段650米长的温哥华标志性海堤。借助于人行道、自行车道和多样化的休息区，公园为人们提供了各式各样、充满生机活力的绿色空间，让游人能够和独具风韵的滨水地区零接触，并为他们提供了温哥华其他地方无法给予的独特而难忘的体验。

　　该滨水设计方案中的美学理念一方面考虑了该地是工业遗址，另一方面考虑了该设计致力于建成一座可持续发展的现代都市公园。为了纪念该地的造船业历史，建造了一系列融入多元设计要素、富有想象力且四通八达的公共空间，其中包括横越小水湾的"独木舟状"的人行天桥。除了公园里随处可见的泄洪管理设施和植被外，小水湾两旁立着成行的定点照明灯，并装有特殊的面板，防止光污染；定制垃圾桶中的压缩器利用太阳能电池板供电，将垃圾压缩，以减少垃圾清运次数。

　　水滨公园一期项目完成了对现有的、贯穿于福溪地区和温哥华市区的滨水漫步道和自行车车道的重建。这一富有创造力的项目表明，环境敏感设计完全适用于高密度的住宅区和商业区，并且若能将历史、社交、娱乐休闲融入其中，就更能凸显其魅力。

| LANDSCAPE ARCHITECT Lango Hansen Landscape Architects | CLIENT City of Oregon City | PHOTOGRAPHER Bruce Forster |

JON STORM PARK
/ OREGON CITY, USA

乔恩斯托姆公园

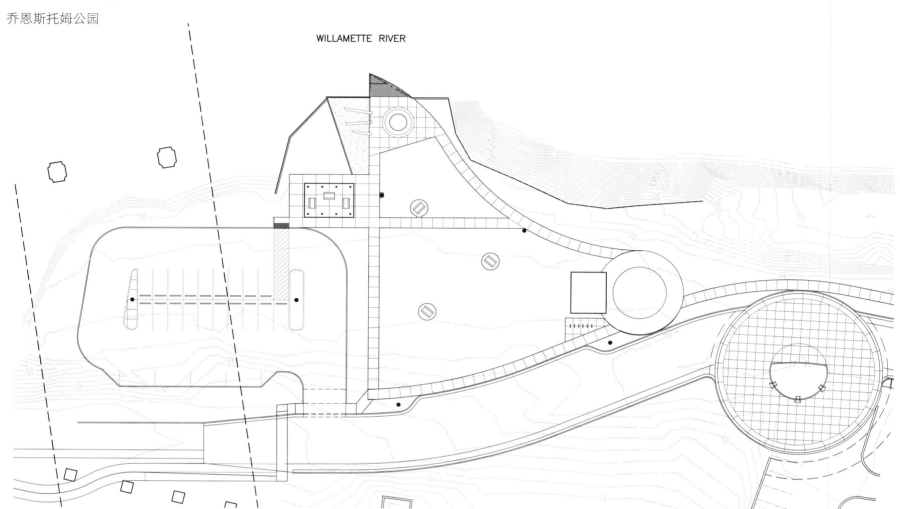

WILLAMETTE RIVER

Jon Storm Park, located in Oregon City, Oregon along the Willamette River near the I-205 overpass, is a 6,070 m² park on the site of a previous log boom operation. The original sheet pile wall that supported a huge crane to lift logs from the log trucks onto rafts below is still in place. The park is a hub of recreational and boating activity with a regional trail that connects to the 99E bike lanes and through Clackamette Park to the Willamette River Promenade. The park and trail design incorporate native plantings, local stone, permeable paving materials, picnic areas, an overlook, interpretive signage and bioswales and detention areas to collect and filter all stormwater on-site. The park functions both as a place to gather for lunch and larger festivals where are held along the waterfront each year.

位于俄勒冈州俄勒冈市的乔恩斯托姆公园坐落于 I-205 天桥附近的威拉米特河沿岸，占地 6,070 平方米。公园的所在地曾是一家兴旺的木材加工厂。当年起重机凭借板桩墙的支撑，将木材从满载木材的卡车上吊起，放到下面的皮筏上。如今，曾经用来支撑巨大起重机的桩板墙仍位于原地。乔恩斯托姆公园是一个休闲娱乐及划船泛舟的好地方。公园的一条小径与 99E 脚踏车车道相连，穿过克莱克米特公园，直通威拉米特河的人行漫步道。公园和步道的设计将很多元素融合在一起：本地植物、当地石材、透水铺装材料、野餐区、眺望台、说明标牌、生态湿地和暴雨降临时能集中和过滤雨水的滞洪区。该公园既可作为午餐聚餐地，也可作为一年一度的海滨盛大节日的举办地。

LANDSCAPE ARCHITECT Häfner/Jiménez **CLIENT** Stadt Staßfurt

STASSFURT
/ STASSFURT AN DER BODE, GERMANY

斯塔斯福特

Stassfurt gained significance with salt mining at the end of the 19th century. Later, the town was to pay a high price for this. Water penetrated the salt mines, dissolved the salt, and the cavities collapsed. The historic city has sunk by approximately 6.30 metres. Within the scope of the International Building Exhibition (IBA) Urban Redevelopment 2010, Stassfurt explored its theme of "Giving up the Old centre" with the aim of giving its lost town centre a new, fit-for-purpose image while at the same time finding strategies for the preservation of memories.

The creation of an artificial lake in the place that was once the site of the medieval centre of Stassfurt commemorates its total collapse due to irresponsible exploitation of potash mines in its subsoil. An area, marks the former church tower, the so-called leaning tower, that was Stassfurt's symbol for over a 500 years.

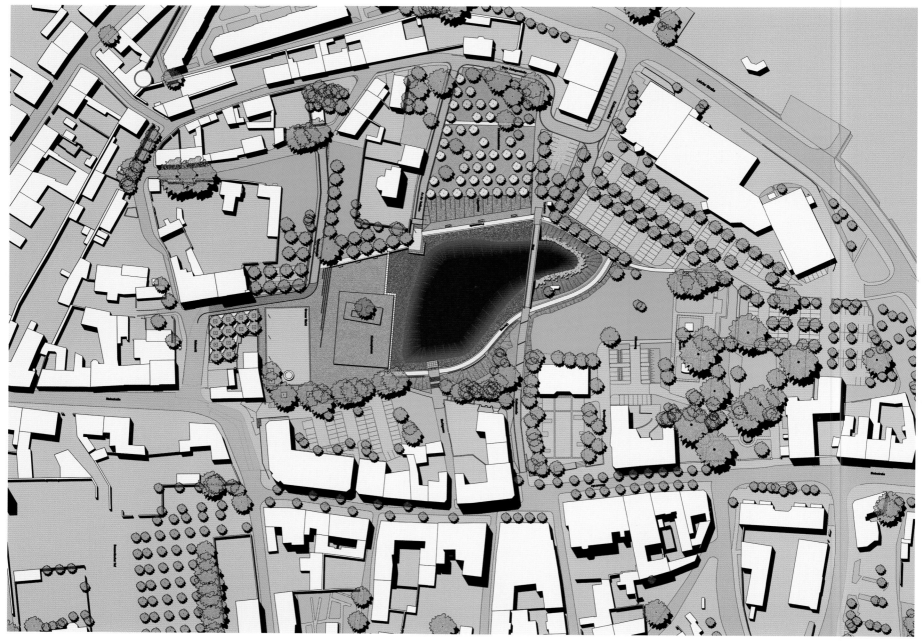

AREA 34,540 m² **PHOTOGRAPHER** Hanns Joosten

ANSICHTEN KIRCHPLATEAU UND ZITAT SCHIEFER TURM

Neupflanzung Sophora japonica

ANSICHT NORD SEE-SEITE

ANSICHT WEST

ANSICHT SÜD MARKT-SEITE

ANSICHT OST

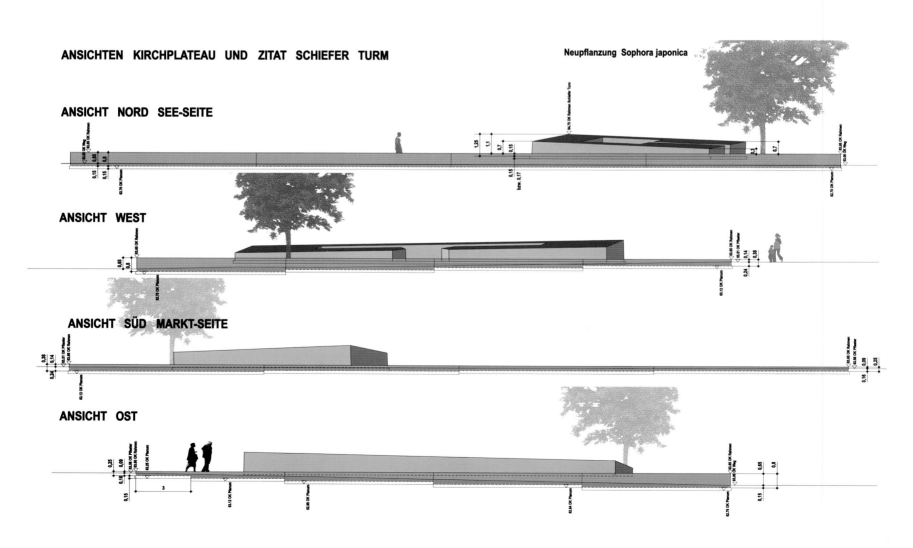

19世纪末,斯塔斯福特因盐矿开采而闻名于世。但是后来,斯塔斯福特市为这一行为付出了高昂的代价。海水淹没了盐矿,盐被溶解,岩盐溶腔坍塌。这一历史名城因此下沉了近6.3米。紧紧抓住国际建筑展2010年旧城改建这一机会,斯塔斯福特进行了主题为"放弃旧中心"的探索,旨在为其所失去的城市中心打造一个合适的全新形象,同时找出保存遗迹的策略。

在曾经作为斯塔斯福特中世纪城市中心的地块上建立起人工湖,以此纪念由于对底土中的钾盐矿不负责任的开发所导致的城市中心坍塌的历史。该区内的教堂塔,就是所谓的斜塔,是具有500多年历史的斯塔斯福特的城市地标。

| LANDSCAPE ARCHITECT HASSELL | CLIENT Australand |

PORT COOGEE REDEVELOPMENT

/ WESTERN AUSTRALIA, AUSTRALIA

库吉港再开发项目

Australand's Port Coogee is one of Australia's biggest coastal renewal projects. The development is a world-class marina and residential estate located 23 kilometres south of Perth. HASSELL is delivering the public domain, including design of the town centre, streetscapes and marina, bridge and lighting infrastructure, and the major realignment of the regional distributor.

The design is inspired by its coastal environment, and cultural and industrial heritage. It expresses these themes through the landforms, plant selection, and through the integration of urban elements that represent the flotsam and jetsam of the shoreline. Port Coogee has been redeveloped respecting the principles of sustainable living with a program that prioritises pedestrian comfort over vehicle traffic and places many key functions within walking distance of the residential development.

In partnership with Australand, HASSELL developed a sustainability approach for the remediation of this post-industrial area that exceeded government standards.

Redundant site materials have been repurposed. A significant bio-regional analysis of the planting structure was undertaken to identify species suited to the sandy alkaline soils and salt-laden winds. Subsequently, plantings were selected for their environmental tolerance and lower fertiliser requirements.

Significant water saving initiatives have been implemented via a comprehensive water management system that converts the abundant groundwater for irrigation. Port Coogee has undergone a distinctly green change that protects and enhances local biodiversity.

Australand began developing Port Coogee in 2006 and HASSELL has played a key role since inception. Partially complete, it features beaches, marinas, waterfront parks, a natural amphitheatre, as well as residential infrastructure. On completion the development will create a new cosmopolitan hub for the region.

AREA 850,000 m² **PHOTOGRAPHER** Peter Bennetts / Australand / HASSELL

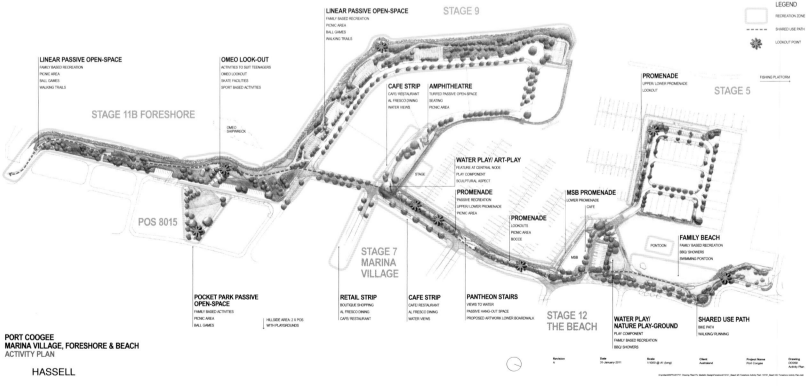

澳洲置地的库吉港再开发项目是澳大利亚最大的滨水重建项目之一。这一开发项目是世界一流的滨水步道和住宅区建设项目之一，距离珀斯23千米。HASSELL得到了这块公有土地设计权，其中包括城市中心设计、街景和步道设计、桥梁和亮化工程及对区域性道路进行重新规划。

该项目的设计灵感来自于其滨海环境、文化遗迹和工业遗迹。该设计通过利用地貌和植被选择来表现这一主题，同时也通过城市元素的综合，例如海边的船只残骸，来表现这一主题。库吉港的再开发工程基于可持续发展的原则：人行步道要优于车行交通，并且很多主要功能区要置于住宅开发区的步行区域。

HASSELL与澳洲置地通力合作，开发出一种可持续性强的方法来修补这个超出政府标准的后工业化区域。多余的建筑材料被重新赋予了新的用途。设计师邀请专业人员进行了一项重要的区域性植物结构分析，以确定哪些植物适于种植在沙碱地上，并能抵抗含盐的海风。随后，设计师按照植物的环境适应能力和低肥需求来挑选植被。

设计师通过装备一套完整的水处理系统来实施意义重大的节水行动。这套系统可以将大量的地下水转化为灌溉水。库吉港切实地进行了一次绿色改革，保护并加强了当地的生物多样性。

澳洲置地于2006年开始库吉港的建设，HASSELL从建设初期就承担着重要的角色。届时已完成了海滩、码头、水滨公园、一个天然圆形剧场和居住区的基础设施建设。工程全部竣工后，这里将成为一个世界性的海港新区。

DIAMOND TEAGUE PARK
/ WASHINGTON DC, USA

钻石提格公园

LANDSCAPE ARCHITECT Jonathan Fitch (Principal-in-Charge) / Lanshing Hwang / Douglas Pardue / Patrick Michaels   **FIRM** Landscape Architecture Bureau

MARINE STRUCTURAL ENGINEER Moffat and Nichol   CIVIL ENGINEER PB Americas   STRUCTURAL ENGINEER Robert Silman and Associates

RIPARIAN HORTICULTURE Signature Horticultural Services

1. **BOARDWALK & BIKE PATH**
   + Gateway to Anacostia Riverwalk
   + Separation of pedestrians & bikes

2. **NATIVE RIPARIAN PLANTING**
   + STABILIZE RIVER BANK
   + PROVIDE HABITAT
   + ECC STEWARDSHIP & TRAINING

3. **FLOATING TREATMENT WETLAND**
   + Stormwater treatment @ major CSO
   + Environmental Education

4. **ENVIRONMENTAL PIER**
   + BOAT ACCESS
   + RIVER VIEWS & LOUNGING

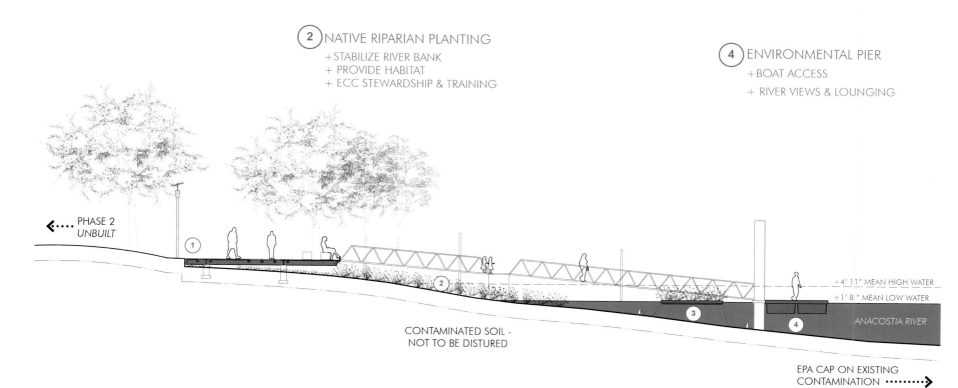

SITE CONTEXT

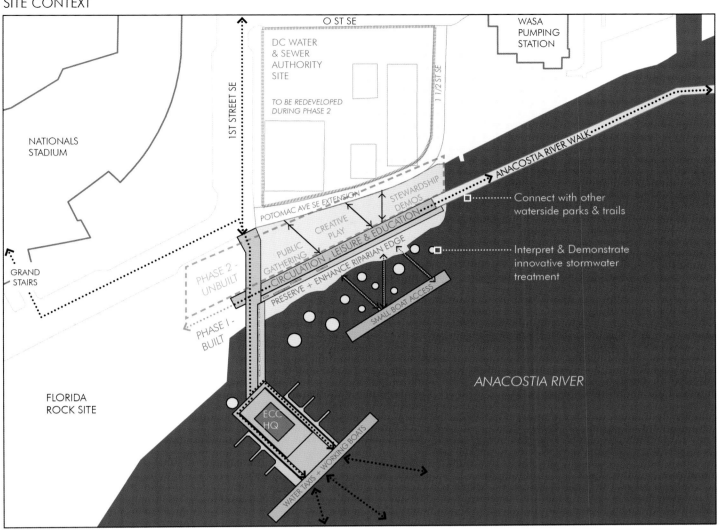

SITE PLAN

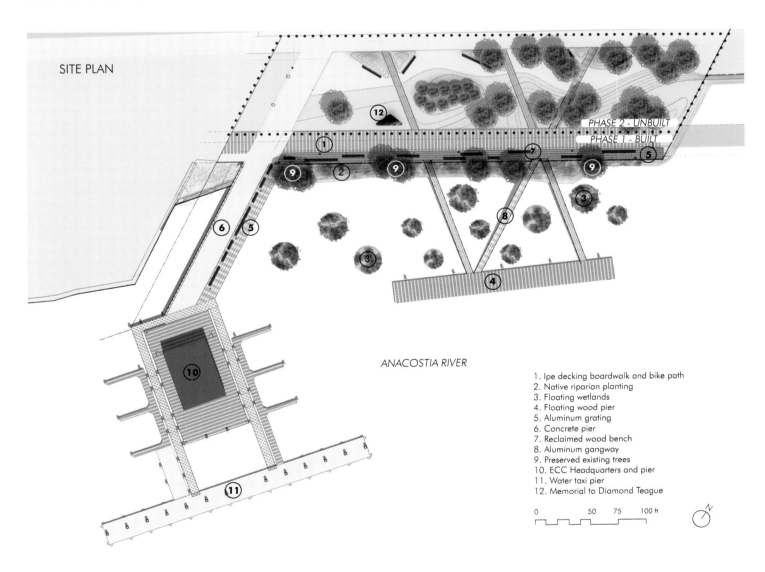

1. Ipe decking boardwalk and bike path
2. Native riparian planting
3. Floating wetlands
4. Floating wood pier
5. Aluminum grating
6. Concrete pier
7. Reclaimed wood bench
8. Aluminum gangway
9. Preserved existing trees
10. ECC Headquarters and pier
11. Water taxi pier
12. Memorial to Diamond Teague

Diamond Teague Park, on the Anacostia River, addresses all of the key issues confronting our cities: environmental degradation and its amelioration, the social divide between rich and poor, economic re-development and enrichment of the sensory world. It is the expression of a clear, simple idea, consistently carried out from concept design through construction. The park consists of a promenade, gangways to a floating pier for kayaks and canoes, the headquarters of an environmental education organization and a pier for water taxis and other large craft. Between the environmental pier and the shore lie floating wetlands--fibrous mats made from recycled plastic in which aquatic plants grow, their roots extending into the water where they filter pollutants. Their circular geometry makes them easily identifiable and reinforces the message that human effort is required to remediate the river. Diamond Teague Park is a case-study in how a simple, urbane, elegant design can, even on a very challenging site, satisfy all of the requirements of use, fulfill an extremely diverse program, and become a well-loved riverside gathering place.

　　钻石提格公园坐落于安那考斯迪亚河上，在那里，我们的城市面临着各种问题：环境的日益退化与改善，贫富差距的增大，经济的复苏和感官世界的富集。公园的设计与建造阐明了一个简单清晰的观念这些清晰的表述和简单的理念都始终贯穿在整个施工过程中。这个公园有一个通往码头的散步长廊——码头那里有很多橡皮船和独木舟，一个环境教育总部，和另外一个有很多水上巴士和其他小船的码头组成。在码头和海滨的中间有一个漂浮的湿地纤维垫，湿地纤维垫是由再生塑料制成，里面可以生长水生植物，这些植物的根可以延伸到水中，起到净化水质的作用。他们的圆形几何使他们很容易被辨别而且同时给人类传达了一个讯息——治理这条河流的必要性。钻石提格公园是一个典型的案例，这个案例阐明了如何创造一个简单而又比较讲究的都市化设计，如何应对各种挑战去创造一个符合各种要求的多样化项目，最终使之变为一个受欢迎的集聚地。

LANDSCAPE ARCHITECT Inspiring Place **TEAM** John Hepper / Jerry de Gryse / Carl Turk / Romilly Davis / Carrie Southern

KANGAROO BAY
/ SOUTH AUSTRALIA, AUSTRALIA

袋鼠湾

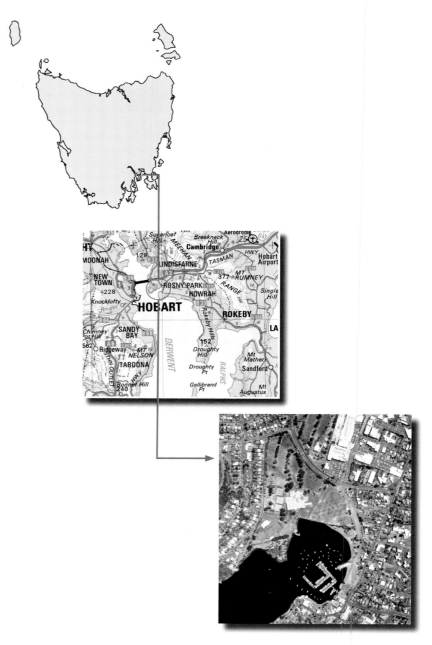

At the start Council was: vexed by how to deal with its significant land holdings and their degraded condition; confronted by inappropriate developer solutions; and pressured by the community to do better.

The Kangaroo Bay Urban Design Strategy (2006) established a vision and mechanisms for the creation of a dynamic mixed-use precinct. The Strategy responded to the qualities of the site, identifying links to nearby shopping areas and suggested the bold move to realign a major road to create a series of substantive commercial/residential development parcels, thus emphasising triple bottom line outcomes.

The Strategy motivated Clarence Council to involve Inspiring Place in the preparation of urban design guidelines (2008) and land re-zoning (2009).

Inspiring Place also worked closely with Council to implement the first two stages of the strategy plan.

• Stage 1 (2009) – A promenade at the Bellerive Yacht Club including a retractable bridge over its slipway – the latter a remarkable achievement negotiated after 20 years of opposition. The promenade is now a civic 'park' inviting use by cyclists and pedestrians who enjoy un-interrupted connection along the foreshore and a ready-made events space. The design is robust, fit for the marine environment, uses recycled materials, and encourages people to stop and engage with the waterfront.

• Stage 2 (2010-2011) - 250m of shared path and foreshore parkland. The design is elegant in its curvilinear form, robust in its materials, fit for purpose in its scale and emphasises sustainability principles.

The Australian Institute of Landscape Architects (Tasmania) recognised the strength of the project in 2011 with an Award for Excellence in Landscape Architecture for Urban Design. The jury stated the project was "an excellent example of the potential for landscape architects to play a key leadership role in visioning and delivery of more sustainable urban design".

ARCHITECTURE / URBAN DESIGN Leigh Woolley   **PHOTOGRAPHER** Jonathan Wherrett / Sharyn Woods

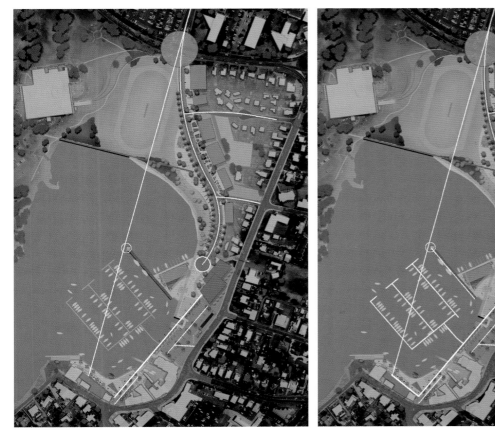

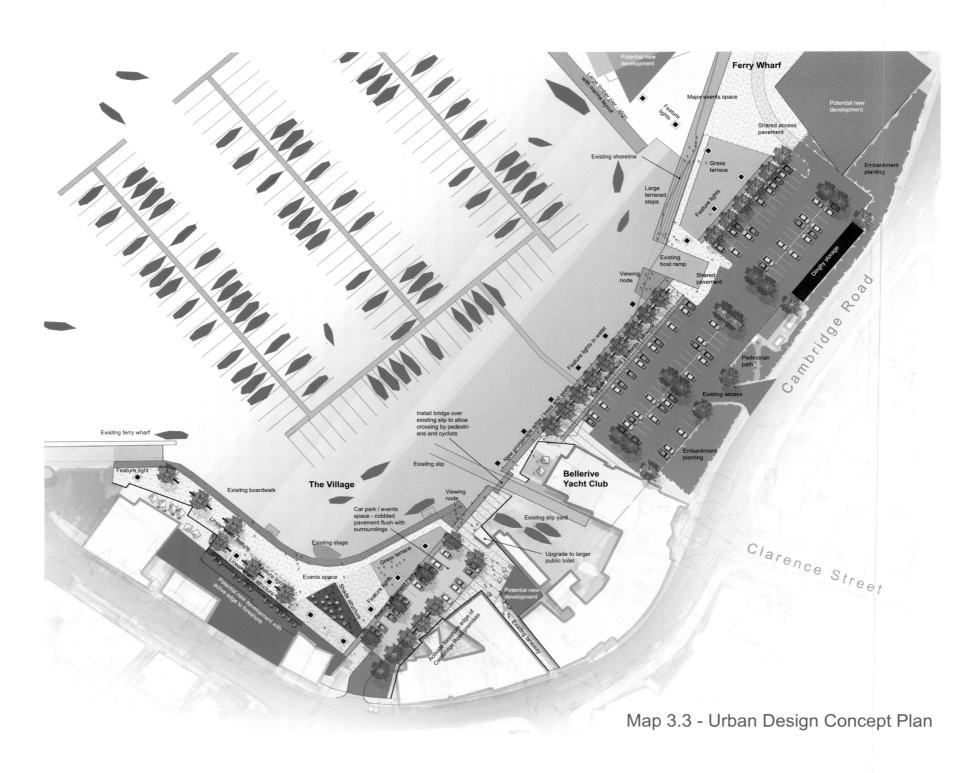

Map 3.3 - Urban Design Concept Plan

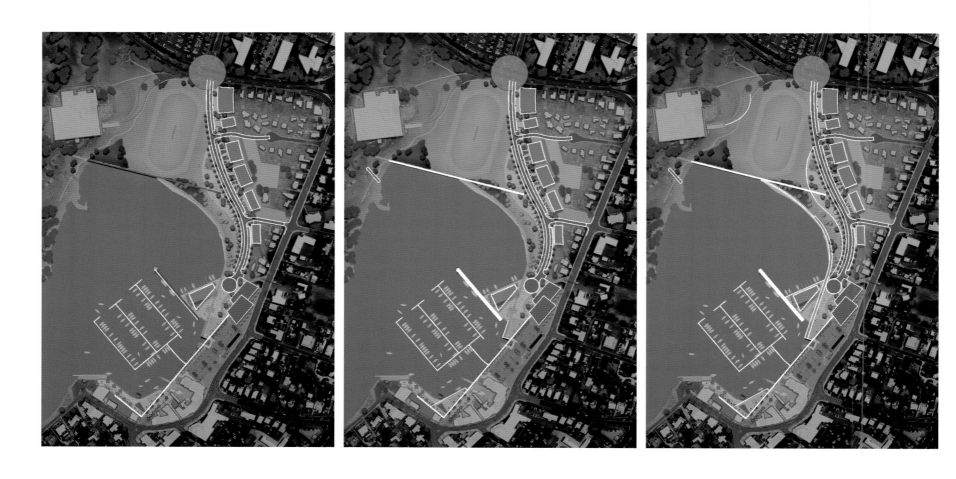

起初，管委会对如何管理所辖重要土地以及如何解决土地的退化问题极为苦恼。管委会还要应对开发公司不恰当的操作，社区也对管委会施压，要求其做出改进。

《袋鼠海湾城市设计策略》（2006）为动态混合用途区的创建确立了愿景和机制。该《策略》对选址的质量做出了规定，明确了与附近购物区的联系，并提议对主干道进行大胆的重新调整，创建出一系列独立的商业／住宅开发分区，从而凸显这三个区域的基本发展方向。

在《策略》的指导下，克拉伦斯管委会将 Inspiring Place 景观事务所列入城市设计导则（2008）和土地区域重新划分决议（2009）的准备工作中来。

Inspiring Place 景观事务所与管委会紧密合作，以便实施前两个阶段的策略规划。

第一阶段（2009）——在贝尔里弗游艇俱乐部旁建造人行步道，包括俱乐部船台上方的可伸缩桥。可伸缩桥是历经了 20 年的对立和协商后所取得的一项非凡成就。现在，人行步道是一座市民"公园"，吸引着自行车爱好者和行人前来，他们喜欢这绵延的海岸交通线和现成的活动空间。本设计稳健，适合海洋环境，使用可循环材料，并且鼓励人们驻足，前来感受海滨美景。

第二阶段（2010-2011）——建造 250 米长的共享通路和海滩公园。该设计的曲线形式优美，材料坚固，符合其规模上的目标，并强调可持续性原则。

澳大利亚景观设计师学会（位于塔斯马尼亚）2011 年对该项目的优势予以认可，并授予城市设计优秀景观建筑奖。评审团称该项目是"证明景观建筑师在规划和开发更具可持续发展的城市设计中发挥潜力并起到关键性领导作用的极好范例"。

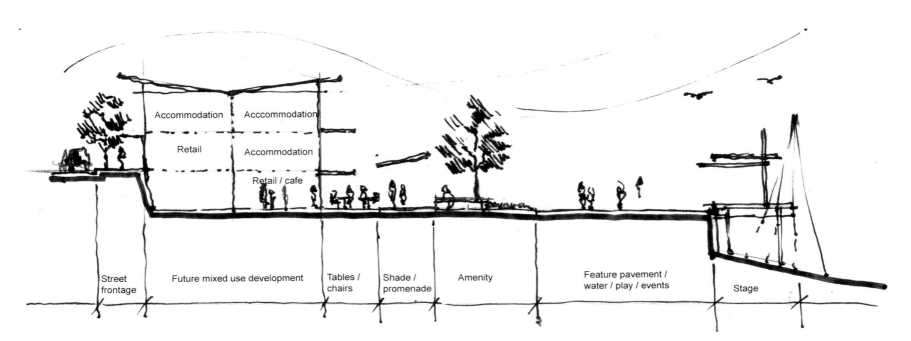

Figure 3.10. Notional Section at Bellerive Village

MASTER PLAN - FEB 2011 - DRAFT
Scale 1:500

Kangaroo Bay

Future Development Site - Ferry Wharf

Labels (clockwise/top-down):
- Bridge - recycled plastic
- Potential landscape terrace seating to be investigated
- Water sensitive urban design 'Education Garden'
- Paint sewer pipeline
- 3.6m pathway extension through to Rosny College. Material to be determined. Maintain constant surface level of path
- Rotated running track alignment.
- Potential additional parking
- Shelter / pavilion
- New rock batter to edge.
- Pedestrian scale pole lighting to pathway on 15m centres
- Mass native plantings.
- Norfolk Island Pines
- Relocate little athletics launch pads
- New Recreational Sporting Facilities
- Potential skate park location
- Grass terrace
- 25m BUFFER
- Water sensitive urban design 'rain garden' design for road and hard pavement storm water run off.
- Green open space
- Kangaroo Bay Drive
- New 5m wide promenade walkway (3.6m shared trail, 1.4m furniture seating zone).
- Potential toilet location central to recreation area.
- New rock batter to edge.
- Integrated and functional path network
- Central picnic facilities. Includes shelter, BBQ(s), seating.
- Green open space.
- **Future Development Site**
- Activate / interact with the waters edge. Access via large stairs and disabled ramp.
- Potential Kangaroo Bay Drive road alignment, roundabout and car parking.
- New 3.6m wide promenade walk way
- Park signage.
- New rock batter to edge.
- Alma Road
- Potential sculpture site
- Regional scale Adventure Playground.
- Viewing platform.
- Kangaroo Bay Drive
- Bus parking.
- **Existing Residential**
- Pickup / drop off parking.
- Turning facility for bus.

INSPIRING PLACE

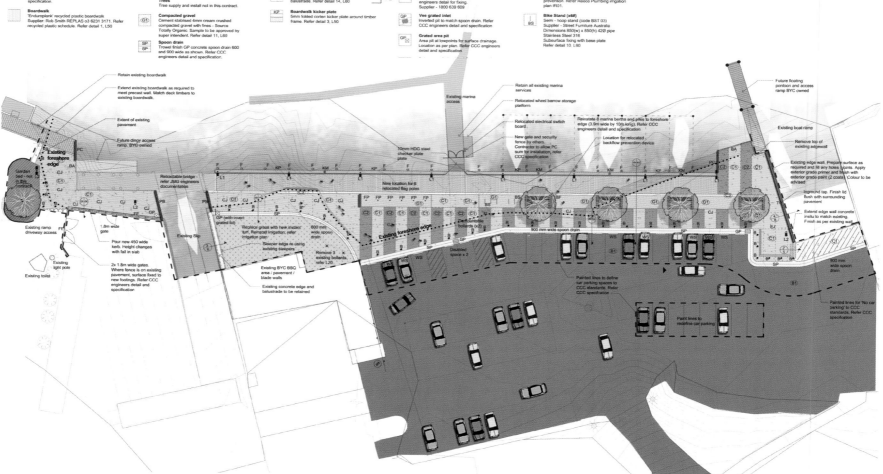

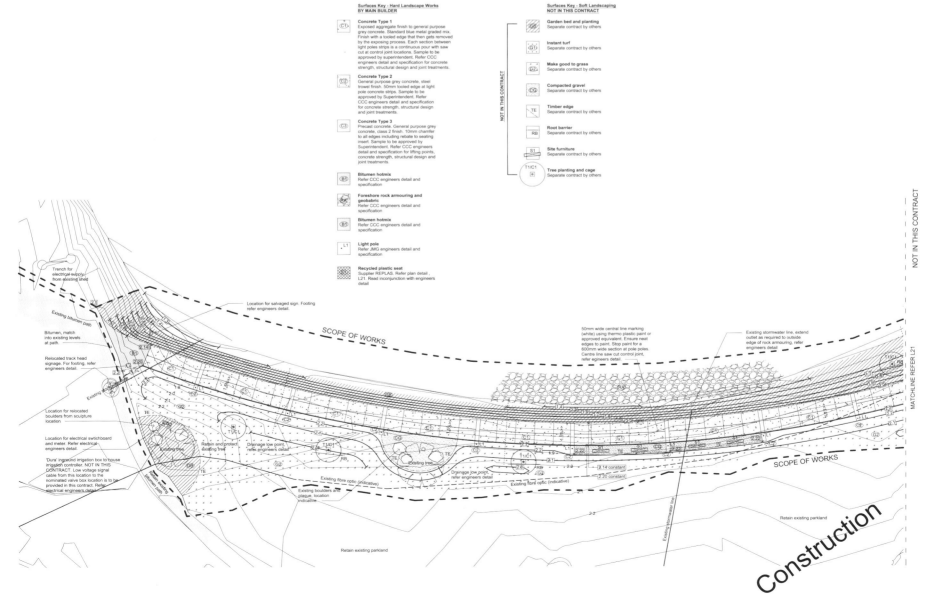

Construction

LANDSCAPE ARCHITECT Kevin Shanley / Scott Slaney   **FIRM** SWA Group   **CLIENTS** Guangdong Nanhai Planning Bureau / Foshan Guicheng Street Authority

NANHAI CITIZEN'S PLAZA AND THOUSAND LANTERN PARK
/ NANHAI, CHINA

南海市民广场和千灯湖公园

In the late 1990's the Shenzhen China Academy of Urban Planning and Design prepared an urban planning document that established a broad framework for the new city center. Upon completion of the planning document SWA was commissioned to provide the urban design master planning for the entire 97-hectare project and subsequently to provide detailed schematic design and construction document review services to implement key infrastructure elements and the 30-hectare "Song of Seven Winds Park," "Lake of 1,000 Lights," and "Citizens Plaza," spaces defined by the master plan.

Restoring a healthy, sustainable, natural ecosystem to site was a prime objective and a driving rationale for the design of park spaces. Islands and wetlands have been created to promote aquatic life. The "White Egret Garden" was developed to attract this native and highly regarded water bird, a Mist Garden and canyon was created to provide a cool and unique experience, and a Flower Maze allows Nanhai to exhibit one of its great skills, the propagation and care of flowers and herbs.

SWA's Song of Seven Winds Park is delineated by seven towers along its edge. Each catches the wind and creates beautiful, melodic sounds inspiring the name. A small concession space is located within the base of each tower with an adjoining terrace that provides vistas to the lake and park.

The 4.2-hectare "Lake of 1,000 Lights" is the focal point of the park. Its cooling presence creates a comfortable microclimate for park users. Boating opportunities also allow people to have recreation on the water and enjoy the park from that unique vantage point. At night park lights, park architecture, and lanterns suspended from boats reflect in the lake animating the park with light.

The project is continuing with additional phases to link the space by water, parks and plazas to additional parts of the city and to create a canal network of vibrant green corridors.

AREA 941,245.5 m²   PHOTOGRAPHER Tom Fox / Jonnu Singleton / SWA

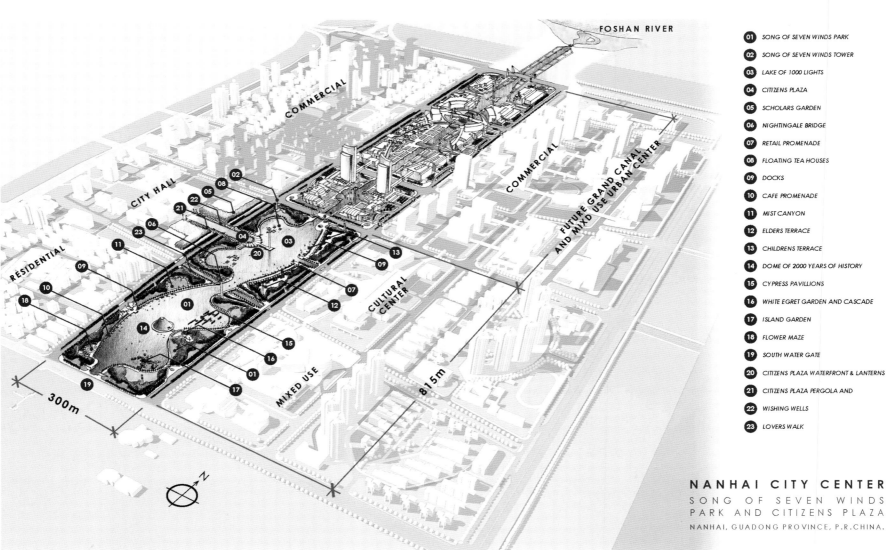

01	SONG OF SEVEN WINDS PARK
02	SONG OF SEVEN WINDS TOWER
03	LAKE OF 1000 LIGHTS
04	CITIZENS PLAZA
05	SCHOLARS GARDEN
06	NIGHTINGALE BRIDGE
07	RETAIL PROMENADE
08	FLOATING TEA HOUSES
09	DOCKS
10	CAFE PROMENADE
11	MIST CANYON
12	ELDERS TERRACE
13	CHILDRENS TERRACE
14	DOME OF 2000 YEARS OF HISTORY
15	CYPRESS PAVILLIONS
16	WHITE EGRET GARDEN AND CASCADE
17	ISLAND GARDEN
18	FLOWER MAZE
19	SOUTH WATER GATE
20	CITIZENS PLAZA WATERFRONT & LANTERNS
21	CITIZENS PLAZA PERGOLA AND
22	WISHING WELLS
23	LOVERS WALK

NANHAI CITY CENTER
SONG OF SEVEN WINDS
PARK AND CITIZENS PLAZA
NANHAI, GUADONG PROVINCE, P.R.CHINA.

20世纪90年代末，中国城市规划设计研究院深圳分院着手制定城市规划，为建造新的城市中心构建总体规划框架。规划完成后即委托美国SWA景观设计事务所负责占地97公顷的工程总体规划。SWA景观设计事务所还要提供具体的方案设计及审查施工文件的服务，以确保建好基础设施以及总体规划中提到的几处景观：占地30公顷的七凤之音公园、千灯湖以及市民广场。

公园规划设计的主要目标和指导原则是恢复当地健康、可持续、且自然的生态环境。岛屿和湿地用来改善水生生物的生存环境；设立白鹭园以吸引这种珍贵的当地水鸟；雾园和峡谷旨在给人们提供一种独特的清凉体验；而花海迷宫则展现了南海人的一大技艺，即繁殖培育鲜花和草本植物。

SWA景观设计事务所在七凤之音公园的边缘位置设计了7座高塔，勾勒起整个公园的轮廓。风吹进塔内会响起优美、悦耳的声音，七凤之音公园因此得名。每座塔的底层都有一间特许经营的小店铺，店铺外面是露天平台，可供游客欣赏湖景和公园景观。

占地4.2公顷的千灯湖是公园的焦点所在。园内很凉爽，为游客创造了舒适的微气候环境。人们在此游船泛舟，在水上尽情娱乐，还可以从这得天独厚的位置欣赏整个公园。夜晚，园内的华灯、建筑和挂在船上的灯笼倒映在湖水里，整个公园便成了灯海的世界。

建设还在进行中，还要通过增建河湖、公园和广场，将现有城区与城市的其他区域联系起来，最终形成一个布满河道和绿色走廊的充满活力的城市网。

/ 245

SITE DESIGN & ARCHITECTURE WEISS / MANFREDI Architecture / Landscape / Urbanism   **DESIGN PARTNER** Marion Weiss / Michael A. Manfredi
PROJECT MANAGER Christopher Ballentine   **PROJECT ARCHITECT** Todd Hoehn / Yehre Suh

OLYMPIC SCULPTURE PARK
/ WASHINGTON, USA

奥林匹克雕塑公园

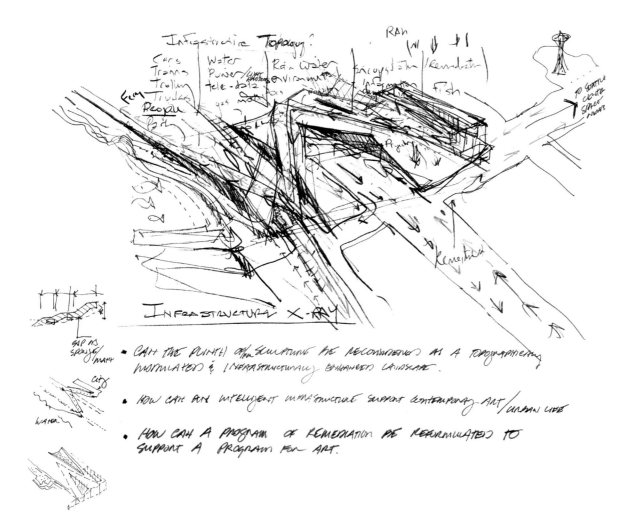

Envisioned as a new urban model for sculpture parks, this project is located on Seattle's last undeveloped waterfront property – an industrial brownfield site sliced by train tracks and an arterial road. The design connects three separate sites with an uninterrupted Z-shaped "green" platform, descending forty feet from the city to the water, capitalizing on views of the skyline and Elliott Bay, and rising over existing infrastructure to reconnect the urban core to the revitalized waterfront.

Winner of an international design competition, the design for the Olympic Sculpture Park capitalizes on the grade change from the top of the site to the water's edge. Planned as a continuous landscape that wanders from the city to the shore line, this Z-shaped hybrid landform provides a new pedestrian infrastructure. The enhanced landform re-establishes the original topography of the site, as it crosses a highway and train tracks, to create a dynamic urban link that makes the waterfront accessible. The main pedestrian route is initiated at an 18,000-square-foot exhibition pavilion and descends as each leg of the path opens to radically different views. The first stretch crosses a highway, offering views of the Olympic Mountains; the second crosses the train tracks, offering views of the city and port; and the last descends to the water. This pedestrian landform now allows free movement between the city's urban center and the restored beaches at the waterfront.

As a "landscape for art", the Olympic Sculpture Park defines a new experience for modern and contemporary art outside the museum walls. The topographically varied park provides diverse settings for sculpture of multiple scales. Deliberately open-ended, the design invites new interpretations of art and environmental engagement, reconnecting the fractured relationships of art, landscape, and urban life.

DESIGN TEAM Patrick Armacost / Michael Blasberg / Beatrice Eleazar / Hamilton Hadden / Mike Harshman / Mustapha Jundi / John Peek / Akari Takebayashi

COMPETITION AND EXHIBITION TEAM Lauren Crahan / Kian Goh / Justin Kwok / Lee Lim / Yehre Suh

CLIENT Seattle Art Museum

PHOTOGRAPHER Benjamin Benschneider / Bruce Moore / Paul Warchol

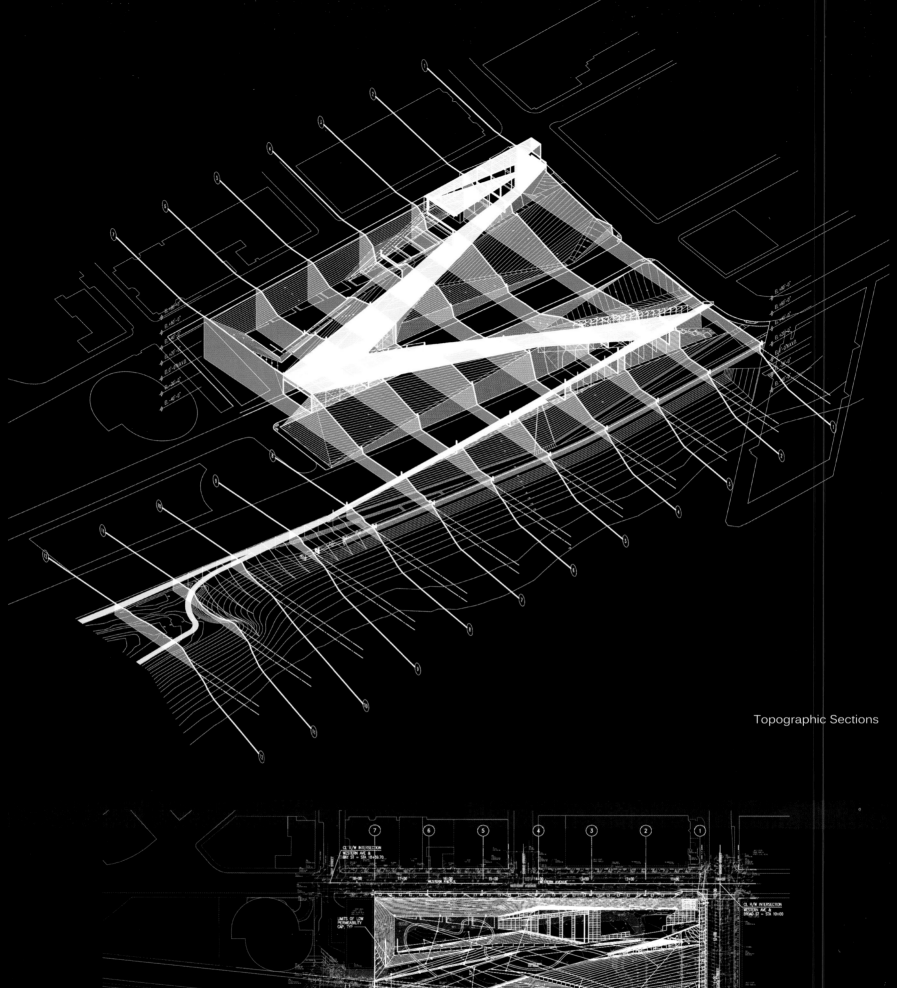

Topographic Sections

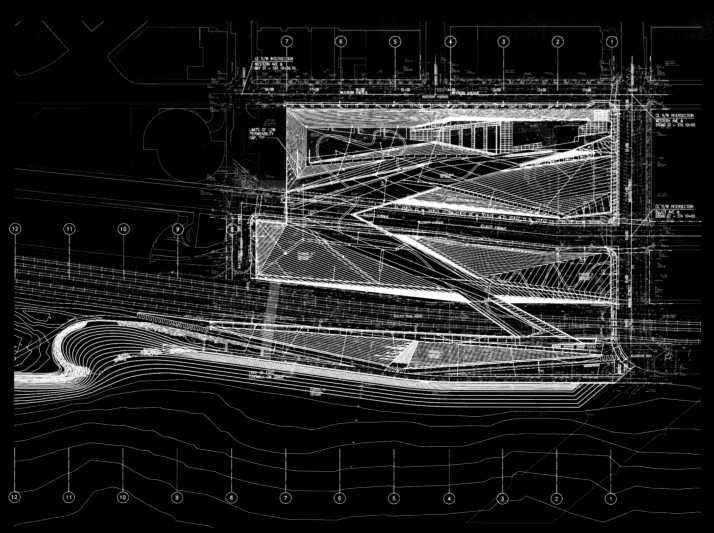

Contour Plan

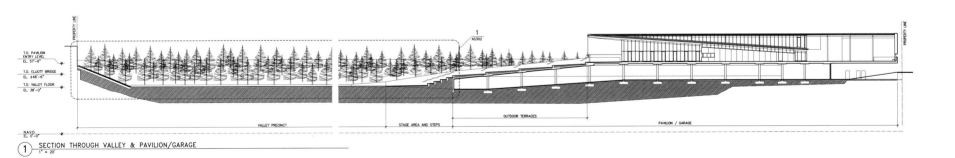

① SECTION THROUGH VALLEY & PAVILION/GARAGE

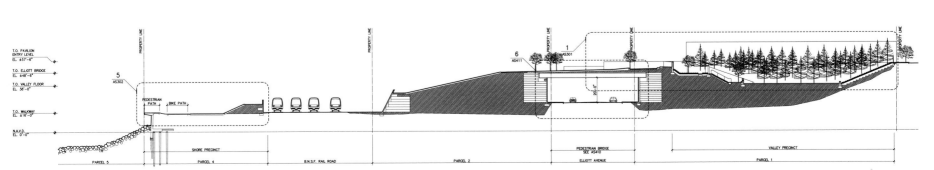

② CROSS SECTION THROUGH ELLIOTT AVENUE BRIDGE

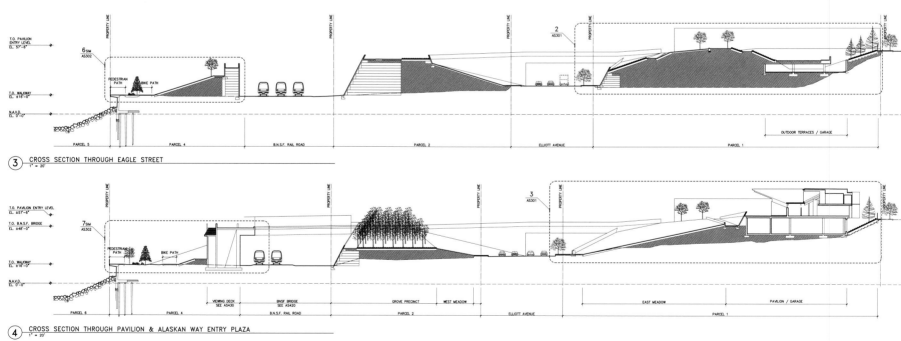

③ CROSS SECTION THROUGH EAGLE STREET

④ CROSS SECTION THROUGH PAVILION & ALASKAN WAY ENTRY PLAZA

　　奥林匹克雕塑公园被看作是新的城市雕塑公园的典范，位于西雅图最后一块未开发的滨海地带——一块被铁道和公路干线分割出来的工业棕地。公园的设计是用连续的 Z 字形 "绿色" 平台将三个被割裂的地块连接起来，平台位于城市和水域之间，其高低落差约有 12 米。因此，从这里可以欣赏天际线和艾略特湾的美景。整个平台建立在原有的地形之上，将城市核心地带与治理后的海滨重新联系起来。

　　作为国际设计大赛的获胜作品，奥林匹克雕塑公园的设计利用了从景观顶端到水滨的层级变化，整个景观是连续的，从城市向海边蜿蜒过去。Z 字形的混合平台提供了一个新的步行通道。该平台重现了原有的地形地貌，并且该通道跨过公路和铁轨，可直达海湾，从而将城市与海滨地区动态地连接起来。主要的步行线路始于约 167 平方米的展馆，并逐渐向下延伸，Z 字形平台的每一段路程都可以看到截然不同的景观。第一段跨过公路，可以领略奥林匹克山的风采；第二段跨过铁轨，可以欣赏城市和海港的美景；最后一段则直通海湾。现在，整个步行平台给人们提供了一个便捷通道，人们可在市中心与治理后的海滩之间自由穿行。

　　作为"艺术景观"，奥林匹克雕塑公园使艺术走出了博物馆，为人们提供了全新的现代艺术体验。公园内多样化的地形满足了多层次雕塑的不同需求。奥林匹克雕塑公园的开放式设计乃设计师故意为之，旨在启发人们对艺术和生态保护做出新的诠释，重新将艺术、景观和城市生活融合在一起。

LANDSCAPE ARCHITECT Place Design Group   **CLIENT** Sunshine Coast Regional Council

MOOLOOLABA FORESHORE STAGE 2A
/ AUSTRALIA
穆卢拉巴前滩重建 2A 期工程

Following a successful Design Development stage, PLACE was commissioned by Maroochy Shire Council to produce construction documentation for the redevelopment of Mooloolaba Esplanade Stage 2A. This beachfront location adjacent to the popular Mooloolaba Surf Club on the Esplanade is a compact foreshore park which experiences concentrated and focused use and demand by residents and visitors alike. The brief was to provide an innovative, creative and functional design in keeping with the established strategic direction and vision for the area. The design needed to capture the essence of Mooloolaba style which required the sensitive manipulation of existing levels and the integration of established trees. Coastal She-Oak, Pandanus and Norfolk Pines were retained as essential character and shade elements throughout the parkland.

The final design required flexibility to find a balance between different uses and their space requirements, including pathways, BBQ areas, lawns, tables, beach access, beach showers and seating coupled with the discrete space available, added to the overall complexity of the project.

The final landscape construction package included plans for demolition and salvage, grading and surface drainage, set-out, surface finishes and details for shelters, stairs, ramps, balustrades and handrails. To protect the existing trees, decks were designed over the existing root zones ensuring a continuity of character and protection of these important assets. The resulting space is heavily utilised by the public and has been a very successful project.

　　一期设计开发顺利结束后,马卢奇郡地方议会委托 PLACE 公司为穆卢拉巴海滨大道重建项目 2A 期工程绘制施工图纸。在这条海滨大道上,有颇具人气的穆卢拉巴冲浪俱乐部,其毗邻海滨区域是一个面积很小的海滩公园,居民和游客对这里的利用和需求尤为集中。该项目的任务是提供一份标新立异、充满想象、功能齐全的设计,并与该区域既定的战略规划方向和构想一致。设计应把握穆卢拉巴整体风格的精髓所在,这就要求对现有状况进行灵活调整并将原有树木融入其中。沿岸的木麻黄、露兜树和诺福克松树作为景区一大重要特色被保留了下来,为整个风景区提供阴凉。

　　最终设计需要灵活安排各种功能用地,权衡其空间需求,包括道路、烧烤区、草坪区、摆放桌椅区、海滩入口、海滩浴场、公共座椅以及被分割开的其他可利用空间,这就使整个项目变得更加复杂。

　　景观建设的后期设计方案涵盖方方面面,包括拆迁和补助计划、平整道路、路面排水、设施陈列、表面抛光处理、林荫区、台阶、坡道、栏杆和扶手等细节的设置等等。为了保护原有树木,在其根部区域设置了围板,这样不仅保持了景观特色的一致性,也保护了景区的重要财产。建成的景区设施被公众广为利用,该项目也获得了巨大的成功。

LANDSCAPE ARCHITECT Taylor Cullity Lethlean / Wraight + Associates

NORTH WHARF PROMENADE, JELLICOE STREET AND SILO PARK
/ AUCKLAND WATERFRONT, NEW ZEALAND

新西兰杰利科北部码头漫步长廊，杰利科大道和筒仓公园

CLIENT Waterfront Auckland (formerly Sea & City) **AREA** 36,422 m² **PHOTOGRAPHER** Simon Devitt / Jonny Davis (Moonlight Silo image)

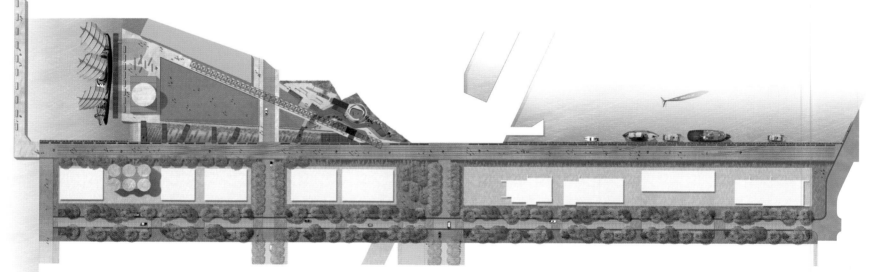

Design Development Plan

The first catalytic projects of this redevelopment are public spaces centered on Jellicoe Street, North Wharf and Silo Park. These spaces promote an alternative design approach to the typical erasure of waterfront memory. Here, friction is encouraged, smelly fish are the attraction, rust, grit and patina are embraced and derelict artefacts are reprogrammed.

Jellicoe Harbour runs parallel to North Wharf Promenade and is used for a diverse array of purposes, including container shipping, ferry services and commercial fishing. Previously, these activities were conducted out of the public eye. Now, however, they are part of the public realm experience and integrated as attractions. A grand axis with a pedestrian focus and rich, informal planting, this "boulevard" establishes a new public realm language for Auckland, one that promotes a civic presence with an indigenous character. Through the narrowing of the overall road dimensions and the introduction of a new tram network, Jellicoe Street prioritises pedestrian circulation over heavy vehicular traffic. The introduction of rough sawn granite sets, fingers of sub-tropical rain gardens, seating and bicycle lanes encourage a human scale interaction with the street and the adjacent Auckland Fish Market.

Silo Park links North Wharf promenade with marine industries to its west. Silo Park is located on a former cement depot, and a large silo that was once earmarked for removal has been retained, forming an iconic landmark. The park now plays host to a range of public functions as passive recreation, events space, youth precinct and weekend market.

Adjacent to Silo Park is a large bio-retention wetland that collects stormwater from the wider site. Indigenous planting references the site's conditions pre-industry. The planting selection references an important element of Maori culture - the viewing of a green edge when approaching land.

By embracing a narrative of place, Jellicoe Street, North Wharf Promenade and Silo Park offer a public realm experience that is both contemporary and sensitive to the site's history, culture and climate.

Silo Section

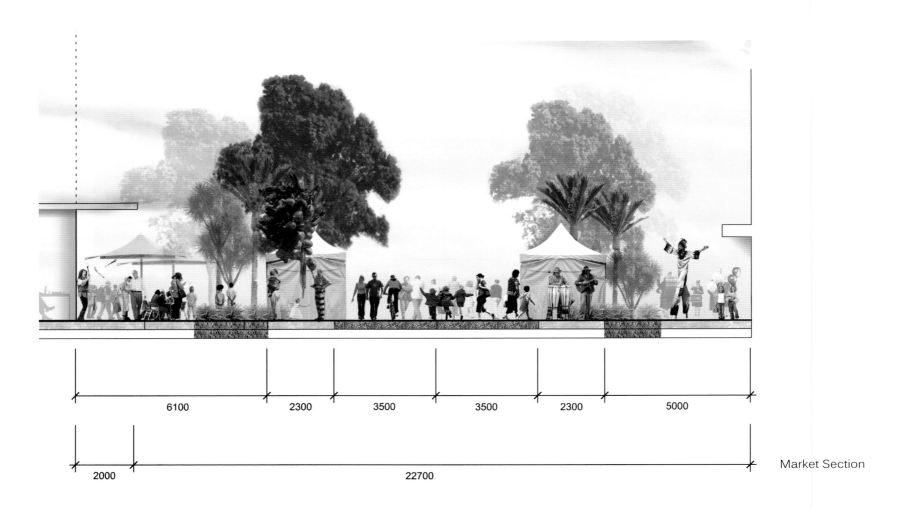

Market Section

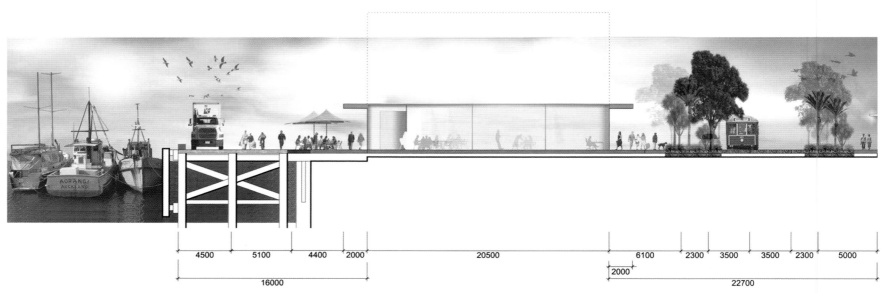

New Jellicoe And North Wharf Section

该项目位于以新西兰杰利科大道、北部码头和筒仓公园为中心的公共用地，是重建开发工程中第一个起到促进作用的项目，催生了一种颠覆人们对海滨传统印象的独特设计方案。这里的设计独树一帜：鱼市成了吸引游客的景致；铁锈、粗砂以及铜绿都应用到设计中；就连被遗弃的人工制品也被重新雕饰。

平行于杰利科北部码头漫步长廊的杰利科港用途广泛，包括集装箱货运、轮渡服务和商业捕鱼。重新整修之前，这些活动已经淡出了公众的视野。现如今，却已成为公共领域体验的一部分，并且极具吸引力。这条林荫大道形成了巨大的轴线，主要为行人使用，道旁栽种着一些茂盛的常见植物，为奥克兰提供了一个全新的公共空间风格，即利用本土特色提升市民的参与度。通过减小道路宽度并引进新的电车网络，杰利科大街上行人的通行优先于车辆的通行。项目中使用了粗糙的花岗岩进行道路铺装，修建指状亚热带雨林花园，设置座椅并开通了自行车道。通过这些人性化设计，吸引人们前往漫步长廊和邻近的奥克兰鱼市。

筒仓公园将杰利科北部码头漫步长廊与西部区域的海洋产业联系起来。筒仓公园的位置曾是一座水泥仓库。一个曾经指定搬迁的大筒仓被保留了下来，形成了形象地标。公园如今包含了一系列公共职能区域，如娱乐消遣区、公共活动区、青年人聚集区和周末集市。

与筒仓公园相邻的是一片广袤的生物保留湿地，这里储藏了从更宽阔的地方汇集来的雨水。本土植物对于这一地区工业时期之前的情况起到了参考作用。对植物的挑选也体现出了毛利文化中一个很重要的元素——当人们接近土地的时候，眼前要看到绿色的景致。

通过融合该地区自身的特点，杰利科大道、北部码头漫步长廊和筒仓公园在历史、文化和气候方面为人们提供了公共领域体验的机会。

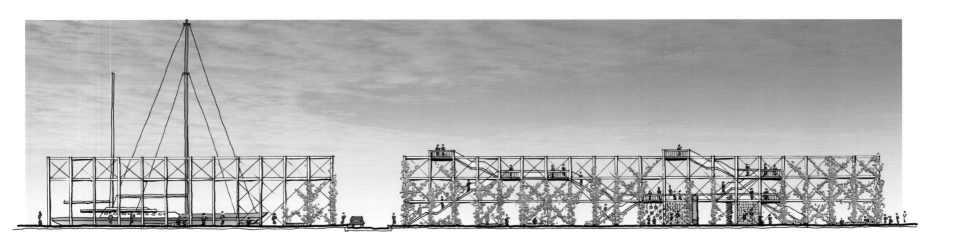

ARCHITECT Emre Arolat Architects   **TEAM** Emre Arolat / Gonca Pasolar / Leyla Kori / Deniz Kösemen / Rıfat Yılmaz / İbrahim Anıl Bicer / Makbule Yıldırım / Yalçın Kaskal / Yadigâr Esen

YALIKAVAK PALMARINA
/ YALIKAVAK, BODRUM, TURKEY

雅勒卡瓦克海滨码头

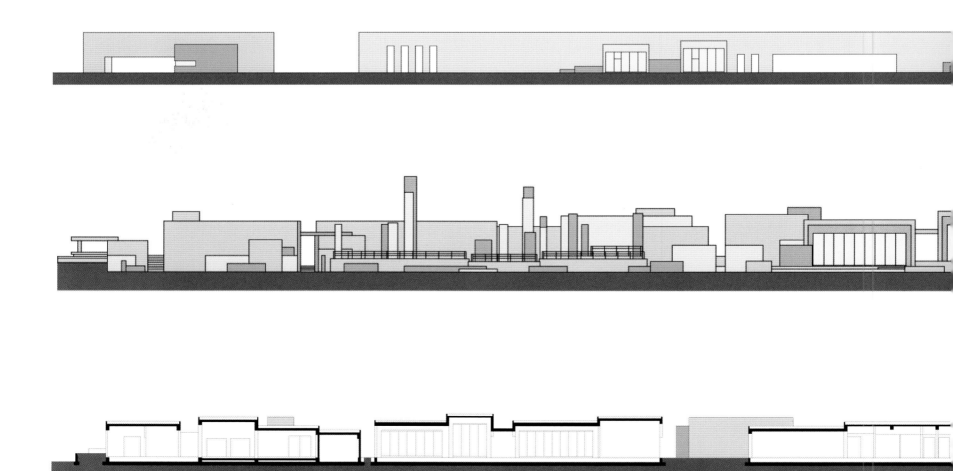

AREA 8,200 m²   **DEVELOPER** Bodrum Yalıkavak Turizm ve Yat Limanı Yatırımcıları ve Ticaret AS

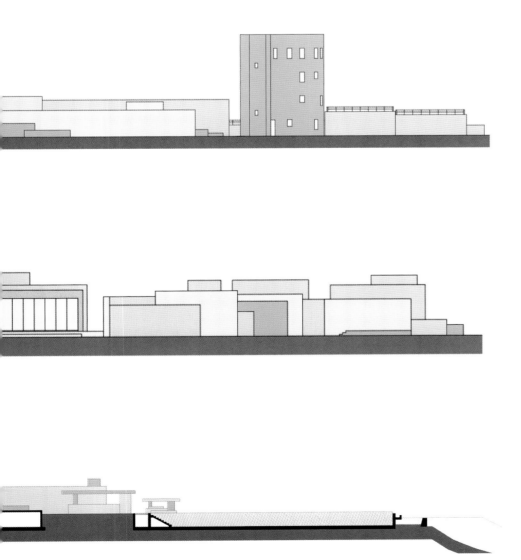

Yalikavak is one of the lagoons on the southwestern coast of Turkey, which is becoming a popular destination for blue voyages along the Turkish Riviera. Unlike its provincial centre Bodrum, which has faced a building boom in 1980s with the increase of touristic activities, Yalikavak is still a relatively calm, smaller scale settlement with its natural landscape.

The project for the extension of the existing marina complex for the use of middle-upper class in Yalikavak has the burden of welcoming a big investment in this area that will also bring its own facilities. The "island" part of Yalikavak Marina, which is the first phase of the complex, is planned to house restaurants, swimming pools, sanitary and mechanical units for the needs of megayachts that will dock in the marina. The main motivation for the design of the "island" was to search for the possibility to reconcile the needs of "outcomers" with the genius loci of Yalikavak as a Mediterranean settlement. Instead of a generic design that can easily become an alienated object for this place, an architecture derived from the local character, interpreted as composition of masses with different heights, merging with landscape and with the sea has emerged as a way to be integrated with the place. Alongside the masses that follows a grid structure in plan, atypical additions such as a lineer wall and a tower accompanies the complex. Following the ancient cities like Kos, Rhodes and Siena, cladded with one material, travertine is used to render the whole complex which is regarding itself as a new-comer, but one of a familiar, not a hard-shell foreigner.

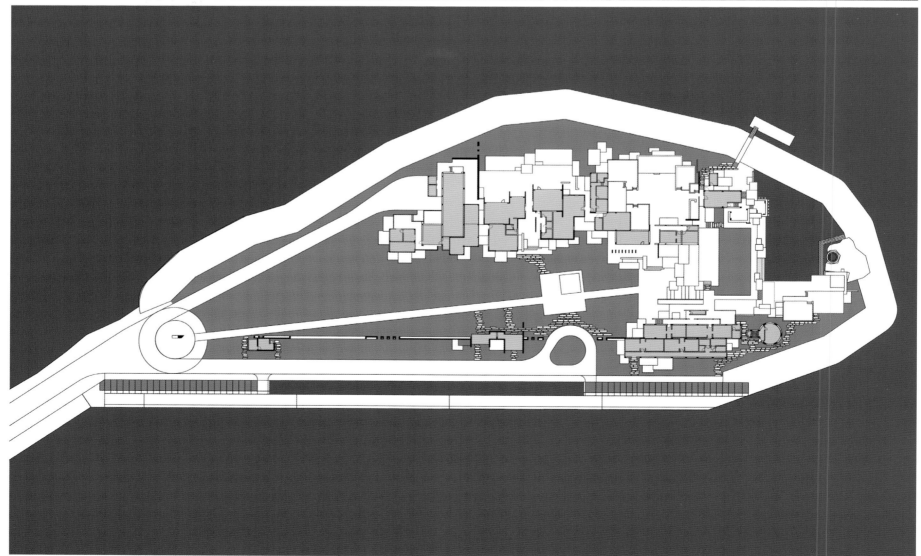

在土耳其的西南海岸上,有着众多的环礁湖,雅勒卡瓦克就是其中之一。现在,雅勒卡瓦克正逐渐成为土耳其里维埃拉海岸上一个备受欢迎的航海旅行圣地。20世纪80年代,由于旅游活动的增多,雅勒卡瓦克所在省份的中心城市——博德鲁姆,经历了一场建设热潮。然而,与博德鲁姆不同,雅勒卡瓦克仍是一个相对安静、规模较小、带有自然景观的宜居地。

为满足雅勒卡瓦克中上层人士的使用需求,码头上现有的综合设施将被扩建。该扩建项目承担着为这一地区吸引重大投资的重任。而引进的重大投资项目也将会带来其自有的建设工程。雅勒卡瓦克码头中的"岛屿"部分,是该综合设施扩建项目中的一期工程。其中包括在码头沿岸建成餐馆、游泳池及卫生与机械部门,以满足停靠在码头的豪华游艇的需要。"岛屿"部分设计的出发点是既能满足观光客的需求,又能保护好雅勒卡瓦克地中海式居住区的地方特色。常规的设计很容易使所建之物与当地的地方特色不符,十分突兀,因此取而代之的是根据当地特色设计的一种建筑风格:众多高度不一的建筑物与自然景观和海洋相融合,这已成为与当地有机结合的一种方式。按计划,众多建筑物将以网格状结构进行建造,旁边则有一些特别附加物,如线型墙体和塔建在综合设施的周围。沿袭了科斯岛、罗兹岛和锡耶纳这些古城的建筑风格,整个建筑群的外部全部采用石灰华材料。既有一些新意,看上去又熟悉和谐,并无突兀之感。

LANDSCAPE ARCHITECT Buro Lubbers **COMMISSIONER** Municipality of Venlo

MAASBOULEVARD VENLO – A NEW URBAN AREA AT RIVER MEUSE IN VENLO
/ VENLO, THE NETHERLANDS

芬洛马斯河大街——芬洛马斯河岸新城区

Since some years the city of Venlo has been working on a major renovation of the urban zone between the inner city and the banks of Meuse river. Buro Lubbers designed the public space of this area. Restructuring and new buildings have transformed the urban wasteland at the Meuse front into a new, dynamic district. Special attention was paid to creating a smooth transition of the historic centre of Venlo to the new area. The connecting link is the quay. Over a length of one kilometre the quay marks the size of the old city. Composed of 3,300 m² broken Belgian limestone the rugged embankment is a solid foundation for existing and new buildings. The quay consists of three levels. The new buildings have retail functions in the plinth and housing at higher floors. At the intermediate level, at the basin, seasonal shops and restaurants with terraces create a lively atmosphere. The lowest level, at the head of the basin, is formed by steps leading towards the water. The peninsula between the basin and the river is designed as a park. The park consists of levels of gently sloping grass areas and small steep edges of basalt boulders. Since these grass slopes are located on different heights, the shape of the island depends on high and low tides. Here the Meuse has free play during the seasons. The Maaspark is part of the regional recreational network, offering a full experience of the Meuse.

COLLABORATOR 3W / Jo Coenen / Arn Meijs   **AREA** 50,000 m²   **PHOTOGRAPHER** Buro Lubbers

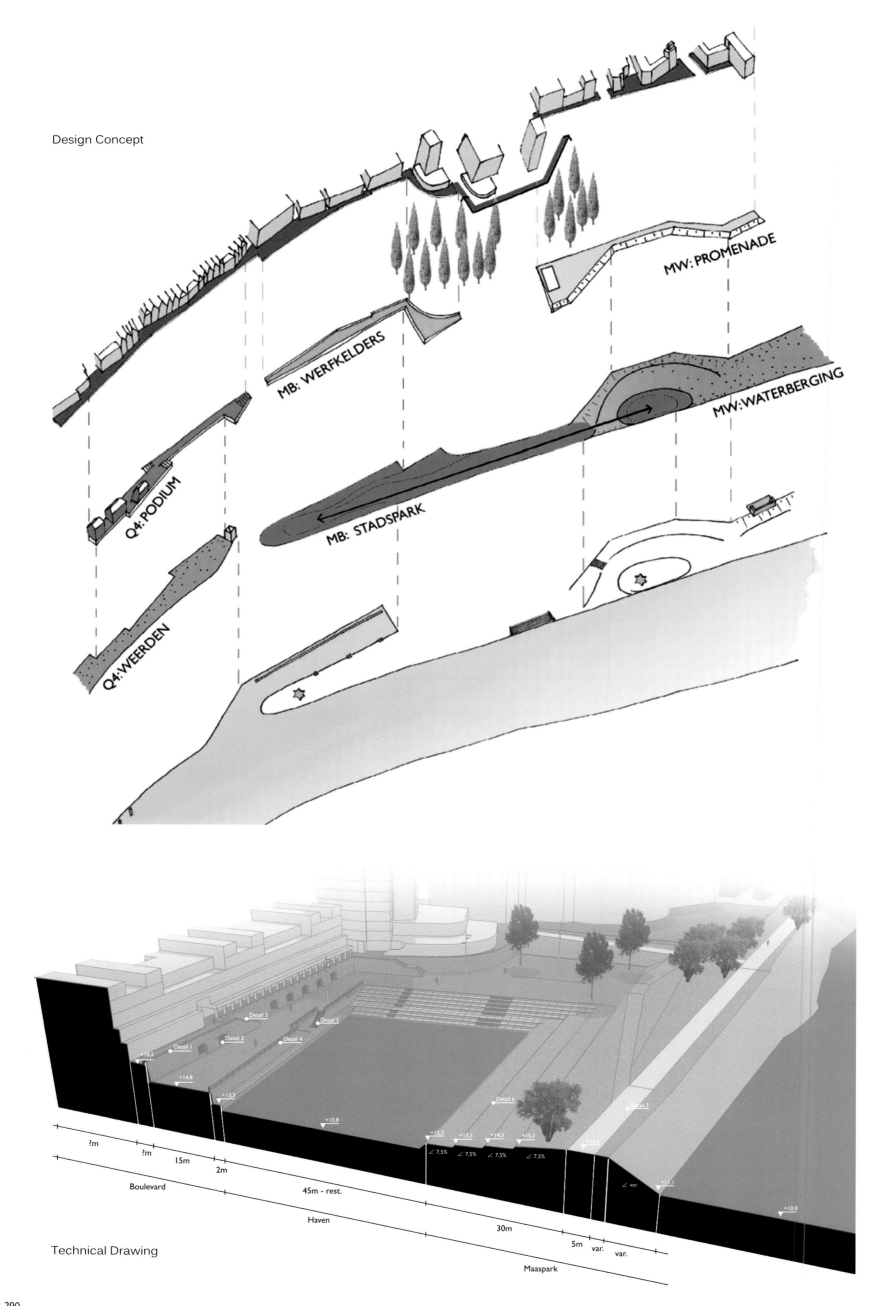

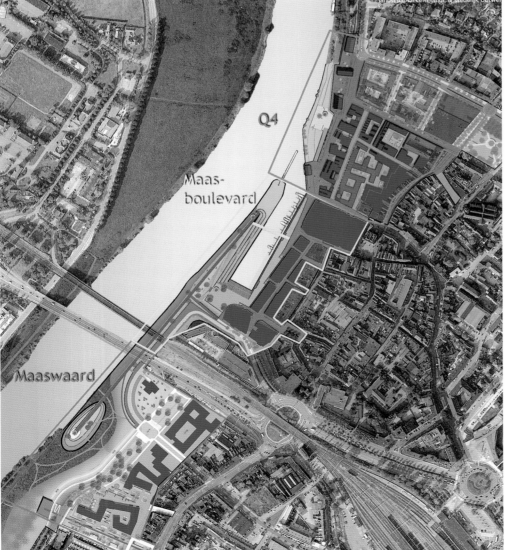

　　近些年来，芬洛市一直在对内城区到马斯河岸中间的城市地带进行重大的整建，由布罗·吕贝尔负责其公共空间的设计。昔日的马斯河河畔荒地经过改建和新建，如今已成为充满活力的新城区。从历史中心到新城区的平稳过渡备受瞩目。而实现过渡很重要的一环便是码头的建设。建成的码头长达1000多米，界定了老城区域。这里凹凸不平的路堤有3300平方米，由比利时石灰石的碎块铺成，为原有建筑和新建筑打下了坚实的基础。码头分为三部分。新建的大楼底层是一些零售店，往上是居民住户。码头的中间层正处内港，有一些季节性的商铺和饭店，都带有露台，很是热闹。在港口的最前端是码头最底部，有几级石阶，走下去便是河水。马斯河同其内港之间的半岛被设计成了一座公园，园内的绿草地高低起伏，坡度却很平缓，并且随处可见带有小陡边的玄武岩鹅卵石。这些草坡所在的地块高度各异，如此一来，潮水的涨落便会改变整个岛屿的轮廓。马斯河一年四季长流不息。马斯河公园是区域内的重要休闲娱乐场所之一，在这里可以尽情领略马斯河的风光。

LANDSCAPE ARCHITECT Sasaki Associates, Inc.   **AREA** 101,171 m²   **PHOTOGRAPHER** Craig Kuhner

WILKES-BARRE LEVEE RAISING - RIVER COMMONS
/ WILKES-BARRE, PENNSYLVANIA

威尔克斯巴里堤防加高河流公地

Like the walled cities of medieval and renaissance Europe which inspired the design of River Commons, Wilkes-Barre also requires a wall for protection. In this case it is to protect the city from the flooding Susquehanna instead of enemy armies. The project goal was to reconnect the people of Wilkes-Barre and the region to the river, incorporating the waterfront into people's daily life and recreation. Two main approaches provide public access to the river. The first was to strategically select two locations to establish 18-meter gateways through the levee to provide on-grade connection to the riverfront. This idea came from a visioning session with the community, led by The Waterfront Center. The gateways also provide visual connection to the river and Kirby Park across the river. The gates will remain open, tucked into the levee embankment, until needed for flood protection. The river floods quickly providing little time between rain in its upper water shed and the project. Rolling flood gates installed at each opening can be closed quickly to protect the city. The design of the gates and their associated plazas celebrates the renewed relationship between the city and the river, its principal geographic feature.

The second approach establishes a wide promenade along the top of the levee. This promenade bridges over each gateway affording continuous visual connection to the river while also linking to a regional greenway system at both ends. This walled levee edge and adjacent promenade minimized the need for a larger earth levee which would have taken more space from the park.

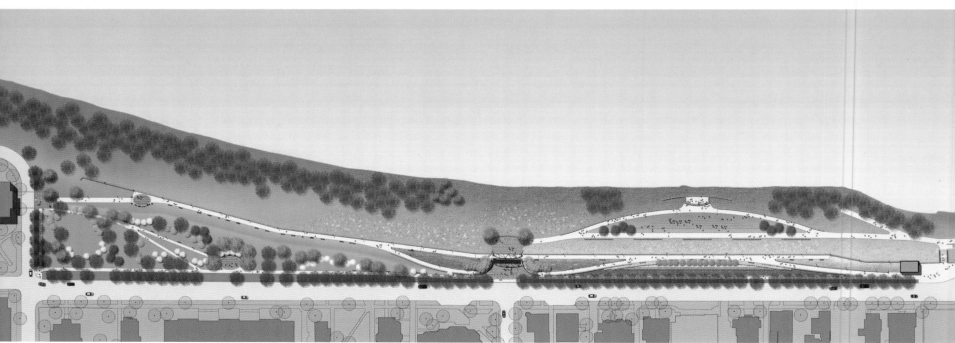

　　中世纪和文艺复兴时期欧洲的古堡城墙给河流公地的设计提供了灵感，同样，威尔克斯巴里市也需要城墙的保护。不过这次不是为了抵御外敌入侵，而是为了保护威尔克斯巴里市免受萨斯奎汉纳河洪水的侵扰。此项计划的目的在于重新构建威尔克斯巴里市人与河流的和谐关系，使人们得以在河滨区域进行正常的工作和娱乐活动。有两条主要的道路可以通往萨斯奎汉纳河。　　设计师在设计第一条道路的时候进行了总体性考量，精心选择了两个位置建起了18米高的出入口。出入口依坝而建，为人们提供了一条整齐平坦的道路通往河滨区域。设计师曾和当地社区共同参加了由海滨中心主办的远景规划会议，这个设计构想就是在那次会议上得出的。人们行走在这条道路上还可以观赏到萨斯奎汉纳河的景色和河对岸的卡比公园。出入口将一直开放，并与堤坝合为一体，若有防汛需要便会关闭。萨斯奎汉纳河的洪水来势迅猛，从上游分水岭突降暴雨到洪水来至泄洪口的时间极短。每一个出入口都装备了旋转式防洪闸，以保证城市在洪水来临时不受侵袭。出入口的设计和相连的水上步道使城市与河流的关系得到更新，毕竟萨斯奎汉纳河是威尔克斯巴里市的地标。

　　第二条道路是在堤坝顶部建设的一条宽敞的散步通道。这条两边筑有防护墙的散步通道绵延不断，将每一个出入口连接起来，水面风光一览无余，在道路两端的特定区域还分布着绿化植被。这样一来，人们不必再建一个更大的土制堤坝，因为一个新的堤坝必将占据公园更多的空间。

LANDSCAPE ARCHITECT Brook McIlroy   CLIENT City of Thunder Bay

PRINCE ARTHUR'S LANDING, THUNDER BAY WATERFRONT
/ THUNDER BAY, ONTARIO, CANADA

桑德贝滨水景观，亚瑟王子码头

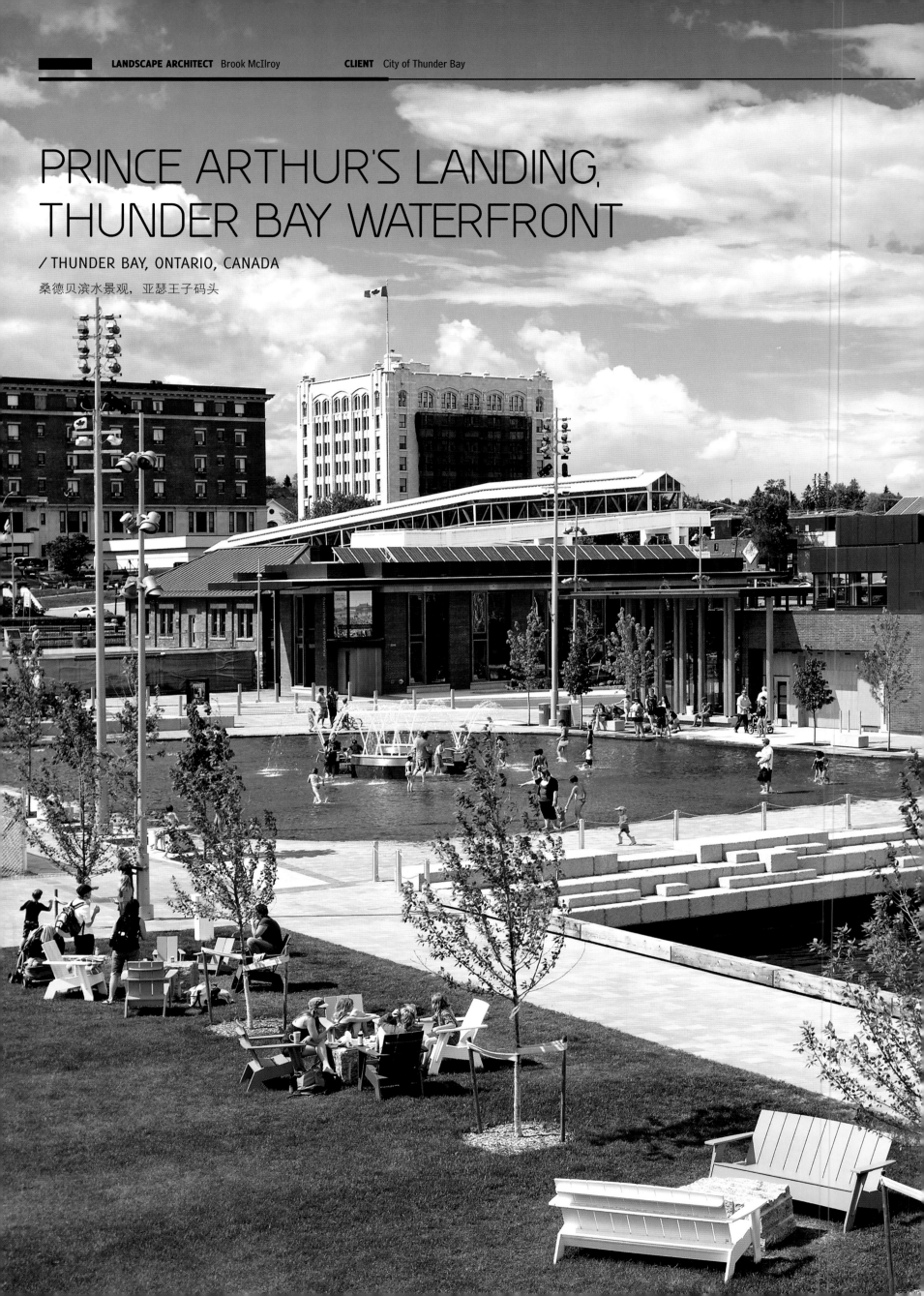

RENDERINGS AND FLOOR PLANS City of Thunder Bay   **PHOTOGRAPHER** David Whittaker

Prince Arthur's Landing transforms the City of Thunder Bay's waterfront into a mixed-use village and animated waterfront park reconnecting the downtown to the shores of Lake Superior. The waterfront opened to the public on December 2011, and has seen record attendance, featuring the opening of several new businesses and over ten design excellence awards.

Commencing in 2006, Thunder Bay commissioned a team led by Brook McIlroy to design its new waterfront that would transform the City's image and quality of life. In 2009, the City was awarded the second largest contribution in Ontario under the National Infrastructure Stimulus Fund (ISF) program. Through a $22 million investment, the City leveraged a comprehensive revitalization project with a construction value of $120 million, composed of $55 million in public sector funding and $65 million private sector investment.

Public elements include:

• Waterfront Park network including dedicated trails and landscaped destinations.

• Water Garden Pavilion – restaurant, event space and support for skating rink / summer splash pool.

• Baggage Building Arts Centre – renovated 1900's heritage structure containing a public art gallery, retail, and artisan studios.

• Market Square and Waterfront Plaza – multi-purpose piazza space.

• Pond Pavilion – support for waterfront recreation activities including rentals.

• Skate Board Park.

• Spirit Garden – outdoor performance/gathering area and rich public park space.

• 276-slip Marina.

Private sector components under construction include: a 150-room Hotel & Conference Centre; two condominium buildings; Market building for retail and office uses; and CN Railway Station renovation to accommodate restaurants and retail.

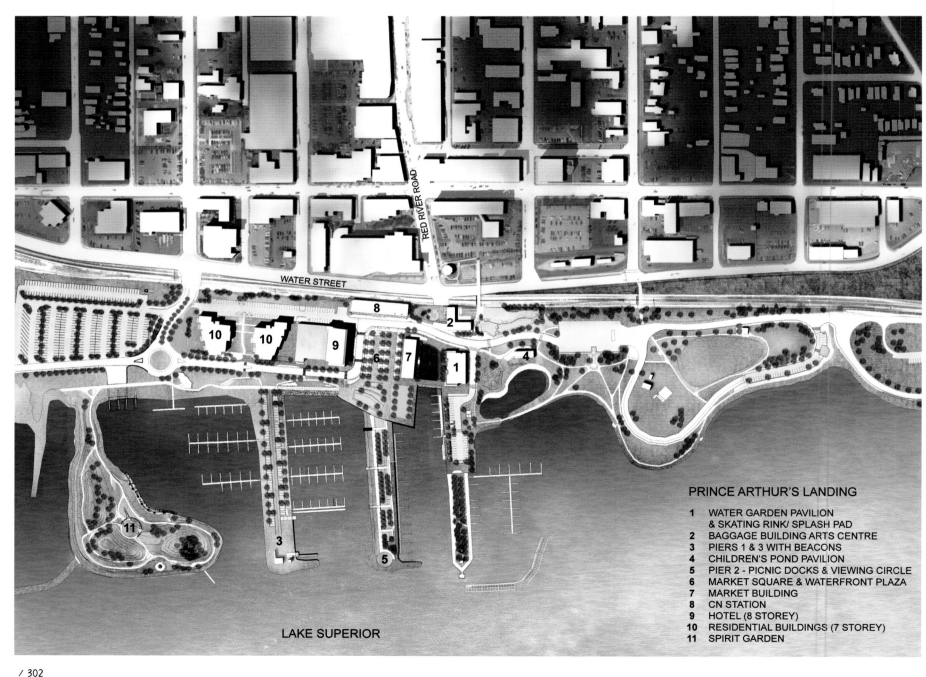

PRINCE ARTHUR'S LANDING

1. WATER GARDEN PAVILION & SKATING RINK/ SPLASH PAD
2. BAGGAGE BUILDING ARTS CENTRE
3. PIERS 1 & 3 WITH BEACONS
4. CHILDREN'S POND PAVILION
5. PIER 2 - PICNIC DOCKS & VIEWING CIRCLE
6. MARKET SQUARE & WATERFRONT PLAZA
7. MARKET BUILDING
8. CN STATION
9. HOTEL (8 STOREY)
10. RESIDENTIAL BUILDINGS (7 STOREY)
11. SPIRIT GARDEN

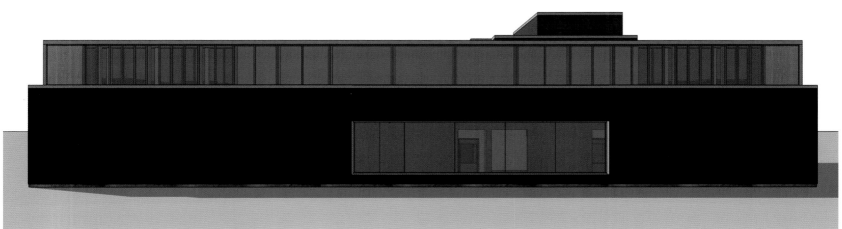

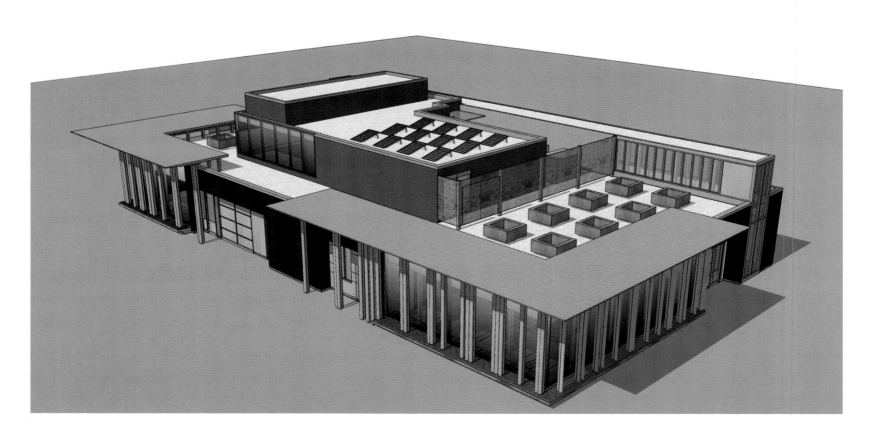

/ 305

　　亚瑟王子码头将桑德贝城的滨水区变成了一个多用途的村落，滨水公园被注入了新的活力并且将市中心与苏必略湖的岸边连接到一起。该滨水景观于 2011 年 12 月向公众开放，吸引了大量游客到此观赏，几种新型商业在这里展开，该景观还赢得了超过十项卓越设计奖。

　　该项目于 2006 年开始启动，桑德贝市政当局委任由布鲁克·麦克劳伊领导的设计团队为该市设计新的滨水景观，以期改善当地的城市印象与生活质量。2009 年，该市被加拿大国家基础设施建设鼓励基金会授予了安大略湖地区最大贡献奖的第二名。该市投资 2200 万，通过举债经营全面开启了这个 1.2 亿的重建项目。举债部分包括 5500 万公共产业基金和 6500 万私人产业投资。

公共建筑包括：
- 包括专用游览路径和观景目的地在内的滨水公园道路网。
- 滨水花园馆：餐厅和可供冬季滑冰和夏季游泳的活动场地。
- 行李楼艺术中心：改建自 20 世纪初的遗留建筑，包括一个公共艺术画廊、零售中心和艺术家画室。
- 自由市场和滨水广场：多用途广场空间。
- 池塘馆：支持滨水娱乐活动，包括出租位置。
- 滑板公园
- 圣灵花园：室外表演、集会区域和巨大的公共绿地空间。
- 276 片码头

在建的私人经营部分包括：一个拥有 150 个房间的宾馆和会议中心、两栋公寓楼和用于零售业和办公的写字楼。此外，CN 火车站也被重建以迎合餐饮业和零售业的需要。

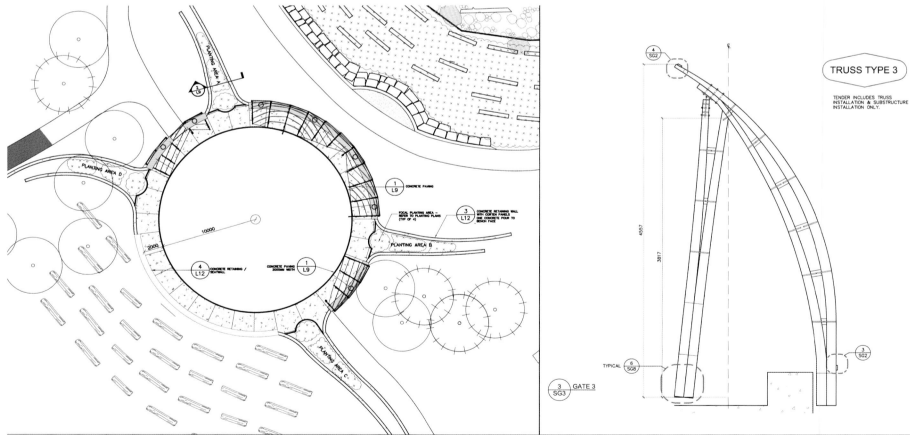

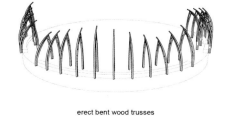
erect bent wood trusses

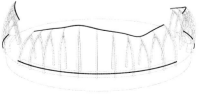
assemble top & bottom rails

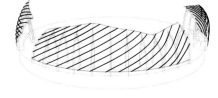
assemble lower substructure

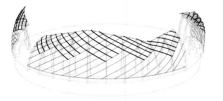
assemble upper substructure

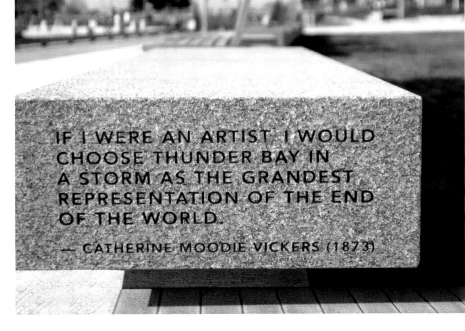

IF I WERE AN ARTIST, I WOULD CHOOSE THUNDER BAY IN A STORM AS THE GRANDEST REPRESENTATION OF THE END OF THE WORLD.

— CATHERINE MOODIE VICKERS (1873)

/ 313

| LANDSCAPE ARCHITECT Grant Associates | ARCHITECT Wilkinson Eyre Architects |

GARDENS BY THE BAY
/ SINGAPORE

滨海湾花园群

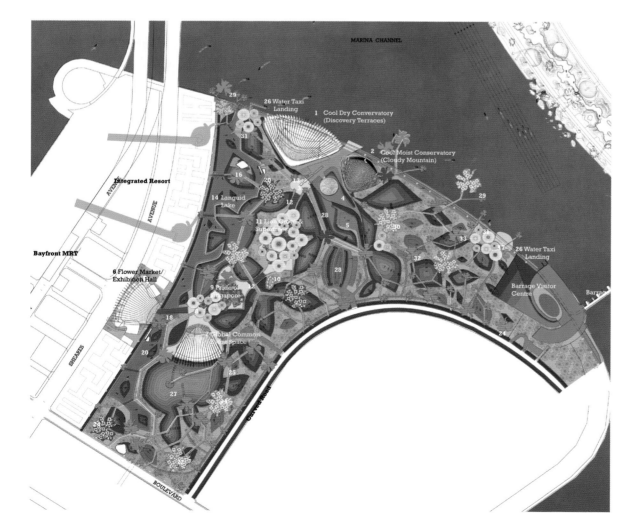

Designed by Grant Associates the Supertrees are unique vertical gardens ranging from 25 to 50 metres in height (equivalent to 9 to 16 storeys), with emphasis placed on the vertical display of tropical flowering climbers, epiphytes and ferns.

There are a total of 18 Supertrees, all located within Bay South at Gardens by the Bay. Out of the 18 Supertrees, 12 are situated in the Supertree Grove while the remaining 6 are placed in clusters of threes near the Arrival Square and Dragonfly Lake. Given the relatively short time span to create a garden from reclaimed land, the Supertrees provide an immediate scale and dimension to the Gardens while marrying the form and function of mature trees. They also create height to balance the current and future tall developments in the Marina Bay area. In the day the Supertrees' large canopies provide shade and shelter. At night, the Supertrees come alive with lighting and projected media created by Lighting Planners Associates.

A 128-metre-long aerial walkway designed by Grant Associates connects the two 42-metre Supertrees in the Supertree Grove to enable visitors to take in a different view of the Gardens from a height of 22 metres. The 50-metre Supertree has a treetop bistro designed by Wilkinson offering a panoramic view of the Gardens and surrounding Marina Bay area.

CLIENT Wilkinson Eyre Architects **AREA** 16,000 m² **PHOTOGRAPHER** Craig Sheppard/Robert Such/Darren Chin

Supertree Grove Plan

Plan of 30m Supertree with PV panels

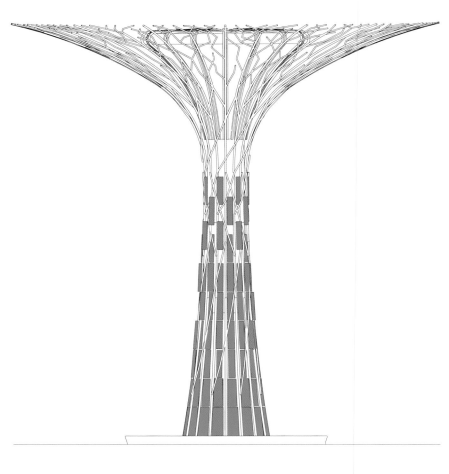

Elevation of 30m Supertree with planting panels

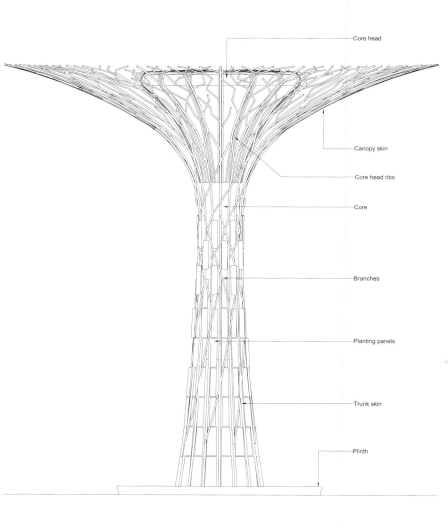

Elevation of 30m Supertree with planting panels

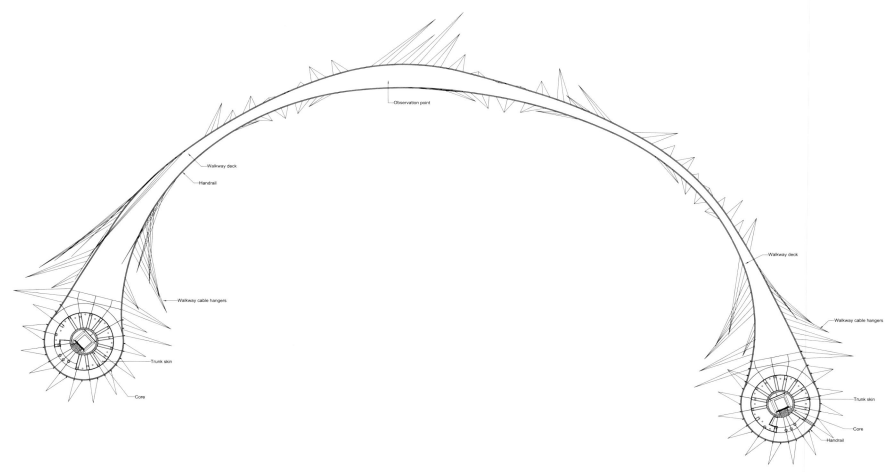

Plan of Aerial Walkway (OCBC Skyway)

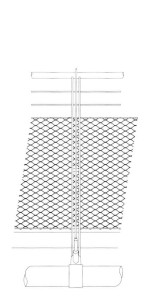
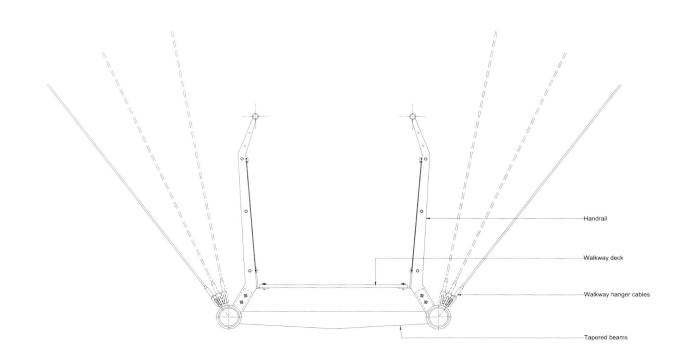

Aerial Walkway (OCBC Skyway) Cross section & handrail elevation

　　由 Grant Associates 景观事务所设计的"擎天大树"是独一无二的垂直花园群，其高度从 25 米到 50 米（相当于 9 到 16 层楼高）不等。该花园群的重点在于垂直展示热带开花攀援植物、附生植物和蕨类植物。

　　花园群共有 18 棵"擎天大树"，均位于滨海湾花园群的海湾南侧。在 18 棵"擎天大树"中，有 12 棵位于"擎天大树林"中，而其余 6 棵则每 3 棵为一丛，坐落在迎宾广场和蜻蜓湖附近。以较短的时间在填海造地内建起的"擎天大树"花园群，将成熟树木的形式和功能结合起来的同时，为各花园提供了相当规模和面积的美化。各花园也留有一定的高度空间，以使滨海湾地区当前和未来在高度上的发展保持平衡。白天，"擎天大树"的巨大树冠提供阴凉和庇护；夜晚，在照明规划师协会制作的灯光和投影媒体的映照下，"擎天大树"又熠熠生辉。

　　一条由 Grant Associates 景观事务所设计的 128 米长的空中通道，将"擎天大树林"中两棵 42 米高的"擎天大树"连接起来，使游客可以从 22 米的高空观赏花园群里的不同景色。一棵 50 米高的"擎天大树"的树顶上建有一个小酒馆，由威尔金森·艾尔设计。在小酒馆中，游客可以观赏到花园群以及滨海湾地区附近的全景。

LANDSCAPE ARCHITECT Taylor Cullity Lethlean   **CLIENT** Land Management Corporation

NEWPORT QUAYS
/ PORT ADELAIDE, SOUTH AUSTRALIA

新港码头

The redevelopment of the Port Adelaide inner harbour is one of South Australia's largest and most significant urban development projects. The planning and design of the first stage of the project was commenced in 2004, based on a previously endorsed masterplan.

Taylor Cullity Lethlean was engaged by the Brookfield Multiplex, Urban Construct joint venture, to undertake urban and landscape design for the first stage including the regeneration of the adjacent portions of Mangrove Park. Through close collaboration with Cox Architects, significant changes were made to the road treatments removing kerbs and creating a more pedestrian friendly environment.

The scope of services included the design of all exterior spaces including the waterfront promenade, outdoor structures and furniture, decking, paving and planting. Working within a tight budgetary framework, care was taken to add detail to furniture and structures wherever possible. Similarly, planting was selected to add patterning, colour and spatial definition to the landscape.

In Mangrove Park at the southern end of the development, extensive soil amelioration and the planting of 37,000 indigenous plants has successfully rehabilitated this formerly degraded section of the park. Consultation with the adjacent school ensured that the park is a venue for their environmental education programs.

The first stage of Newport Quays provides residents and the general public with a readable hierarchy of private, communal and public spaces which contribute to the landscape amenity of the local neighbourhood and broader region. The rehabilitation of Mangrove Park and the design and construction of the first stage of the waterfront promenade with its associated decks, shelters and gathering spaces are particularly important aspects of the project for the enjoyment of the wider community.

PHOTOGRAPHER Ben Wrigley

阿德莱德港内港地区的重建工程是南澳大利亚州规模最大、最重要的城市发展项目之一。2004 年，根据之前审批通过的项目总体规划，该工程第一期规划设计开始动工。

项目设计师 Taylor Cullity Lethlean 受雇于城市建设合资企业——Brookfield Multiplex 公司，负责第一期工程的城市和景观设计，其中包括红树林公园附近地区的重建工作。该公司与 Cox 建筑师事务所密切合作，进行路面处理，移走街头的边石，使道路状况得到了极大改善，为行人创造了更加友好、便利的环境。

项目负责公司的业务范围涵盖所有外部空间设计，包括建造海滨长廊、完善户外建筑和设施、铺设甲板、铺砌道路和种植花草。尽管预算紧张，设计师仍使户外建筑和设施的细节都尽可能完美。同样，植物花草也经过了精挑细选，使景观在图案设计、色彩及空间定义方面更加完善。

在项目南端的红树林公园，设计师对土壤进行了改良，并且种植了 37000 株本土植物，成功地改善了公园原有的退化土地。经与附近学校的协商，公园已成为环境教育项目的定点场所。

新港码头的一期建设为居民和公众提供了层次分明的私有空间、公有空间和公共空间，增加了附近及周边地区的园林风景设施。红树林公园环境的改善、海滨长廊以及相应的露天平台、林荫区以及集会场所的一期设计和建设，都是整个项目尤为重要的部分，为更多的社区提供了休闲娱乐场所。

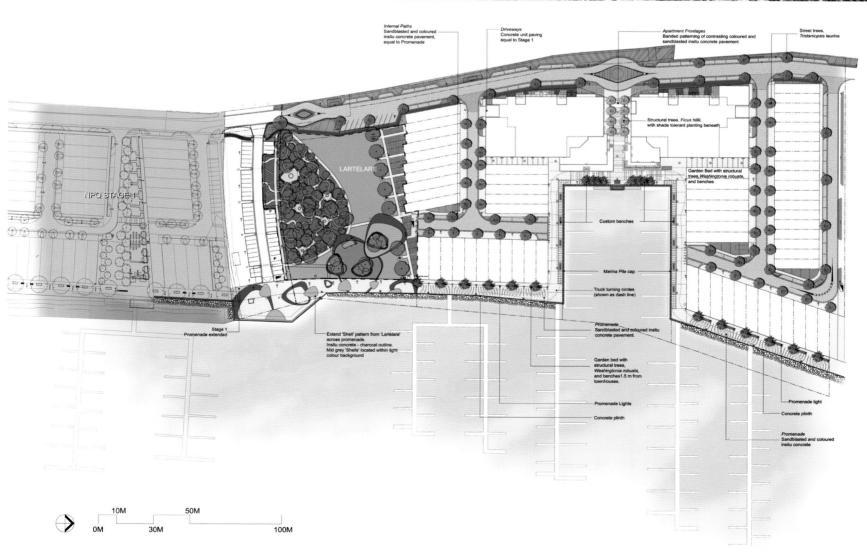

LANDSCAPE ARCHITECT Lucie Bibeau (Chief Urban Planner) / Réal Lestage

CONSORTIUM Daoust Lestage / WAA / Option Aménagement

PROMENADE SAMUEL-DE CHAMPLAIN

/ QUEBEC CITY, CANADA

萨缪尔·德·尚普兰滨水长廊

Quebec City is the oldest city in North America and a UNESCO World Heritage Site.

The Master Plan for the Promenade Samuel-de-Champlain covers more than 10km of public spaces along the majestic St-Lawrence Seaway. It creates a coherent landscape which makes the north shore of the river accessible to pedestrians and cyclists while preserving key industrial amenities that lead to the old historic district.

One of the main features of the proposal is to re-qualify the existing highway which borders the coast into a landscaped, permeable urban boulevard. Drawing on the site's unique history and genius loci, the project uncovers and showcases vestiges of natural and coastal heritage, while balancing luscious greenery with an evocative man-made landscape. The project delicately weaves a sequence of diverse experiences and atmospheres, navigating from the boundless visual expanse of the river while developing tactile and sensory experiences at human scale.

The first phase of the project was completed in 2008, covering a distance of 2.5 km. Four thematic gardens and landscaped spaces are connected by a sinuous and undulating promenade along the river. Each of these dense spaces captures and magnifies the material and poetic qualities of the local coastal environment. They celebrate the river's moods, its mists, winds and sensory pleasures, as well as evoking the memory of former maritime docks.

Thematic gardens line the site, each offering a unique experience influenced by local historic, recreational, ecological and cultural traits. These attraction nodes ensure that the project is vibrant, highly frequented, and integrated into the surrounding communities.

AREA 2.5 km linear park　　　**PHOTOGRAPHER** Vincent Asselin

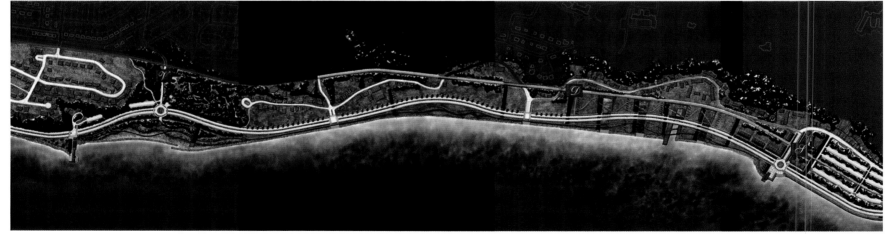

魁北克市是北美最古老的的城市之一，而且被教科文组织列为了世界遗产。

萨缪尔·德·尚普兰滨水长廊覆盖了圣劳伦斯河岸长达10千米的公共区域。在保留通往老城区的一些主要工业设施的基础上，该设计创造了一系列的景观，使北河岸与滨河散步道和单车道相连。

这个设计最主要的特征之一是重新定位现有的滨河大道，将其变为美丽而生态的林荫大道。受当地独特的历史条件和地方特色的启发，该项目充分展现了自然痕迹和河岸遗址的魅力，同时通过极具感召力的人造景观呼应柔和优美的河岸线。该长廊精心地创造了一系列不同的体验以及氛围，让人们在拥有无边无际的视野的同时获得一种可触及的感官体验。

该滨水长廊的第一期在2008年完成，长达2.5千米。将4个主题公园和公共空间由一个蜿蜒波浪起伏的滨河散步道连接，每个景观细节都捕捉到当地河岸环境中富有诗意的材料和质感，并将其放大。景观小品的设置反映了薄雾，海风和海水所带来的感官愉悦以及对船坞区的记忆。

四个主题公园将整个项目联系在一起，依托这一地区特有的历史，娱乐，生态和文化优势，每个公园都给人一种独特的体验。这些优势也使整个项目充满生机，人流量大，并与周边社区融为一体。

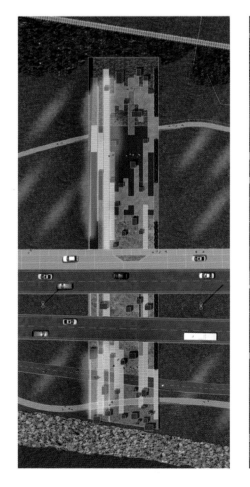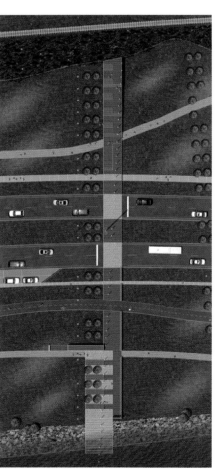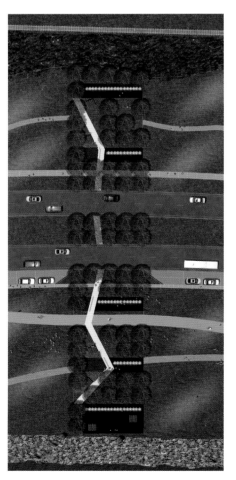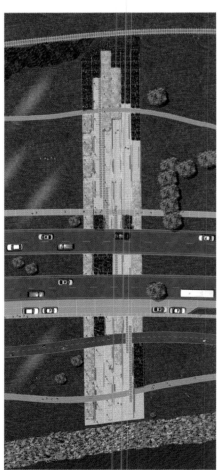

LANDSCAPE ARCHITECT Atelier Corajoud-Salliot-Taborda (HORIZON) / Michel Corajoud Paysagiste / gérant

PROMENADE DES ANGLAIS
/ NICE-PROMENADE DES ANGLAIS

盎格鲁滨海路

The "Promende Des Anglais" (the English Promenade) is registered in the collective consciousness as an international heritage and doubtless as a myth. As such, reorganize it, committed our team to adopt in this design competition, in prerequisite, a real ethics of project. Improve, consolidate certainly but, not to the point, that at the end of this development, we cannot collect any more of its memories in the modified reality. Certain configurations of the « Promenade » have, at least in their spirit, qualities which ought, in our sense, remain intact.

The « Promenade Des Anglais »project extends from the Etats Unies docks up to the of the avenue of the Lantern.

The main guidelines of this project are the following ones:
• Assert the « Promendade Des Anglais » as a world reference
• Restore the link between the City and the sea
• To develop " The Prom " as space of leisure activities and culture

TEAM HORIZON / Claire CORAJOUD / CLARAC / HODEBERT / OGI / MAGOS / ERAMM / ETC **AREA** 200,000 m² **PHOTOGRAPHER** Golem Images Sarl

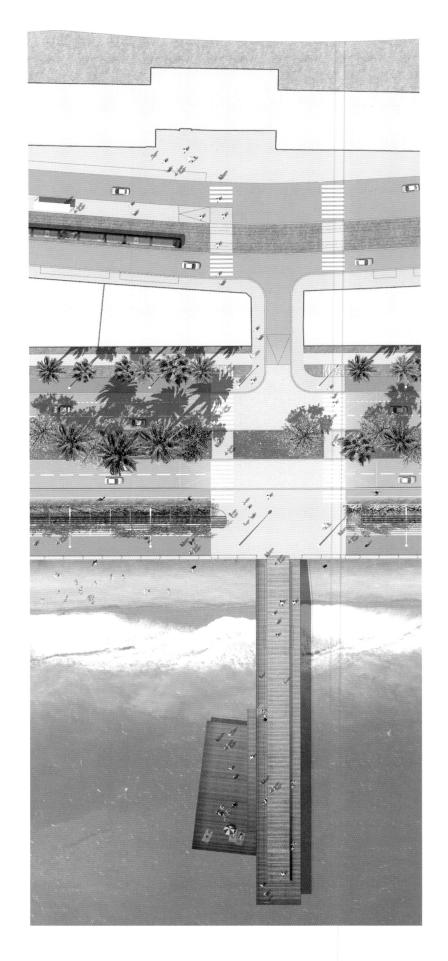

盎格鲁滨海路,也叫"英国滨海路",是公认的世界遗产,也无疑是一个传奇之地。鉴于此,设计团队参加此次大赛时就一致认为,整改绝对是项目进行的前提条件。对滨海路的改进和加强是必要的,但更关键的是,不能在工程结束时将其历史记忆也一并埋葬。滨海步道的某些布局至少在一定风格上是很优秀的,因此,设计团队认为应该保留。

"盎格鲁滨海路"项目的起点为联合船坞,延伸至兰特大街。

项目的主要指导原则如下:

1. 将"盎格鲁滨海路"打造成国际知名地标;
2. 重新把城市与海洋联系起来;
3. 发展"滨海人行道",使其成为休闲活动和文化娱乐场所。

APPENDIX 附 录

ASPECT STUDIOS

ASPECT Studios is a group of design studios united through a philosophy that delivers innovative landscape architecture, urban design and digital technologies.

Since it's beginning, ASPECT Studios has grown to over 55 people throughout their Australian and China studios. They have established a reputation for design-led solutions and are recognised as a company with the capability to deliver world leading design excellence through creative and sustainable projects.

LANGO HANSEN LANDSCAPE ARCHITECTS

Lango Hansen Landscape Architects (LHLA) provides a wide range of services in landscape architecture, planning, and urban design. Over the past nineteen years, the firm's principals have successfully designed public parks, urban plazas, school and university campuses, corporate headquarters, private residences, and public facilities.

LHLA approaches each project as a unique opportunity to develop designs that address the particular character of the site, the specifics of the program, and the needs of individuals and communities. Using a variety of media, such as models, sketches, and computer-aided tools, the firm explores integrated design solutions. With a commitment to detail and craftsmanship, LHLA creates long-lasting designs that express the innate character and value of each landscape.

JONATHAN FITCH

Jonathan Fitch, founder of Landscape Architecture Bureau (LAB), graduated from the University of Michigan with a Master of Landscape Architecture in 1977. Mr. Fitch has extensive experience in the site planning, landscape design, management and implementation of a broad range of projects nationwide and overseas. His work has consistently engaged highly urbanized conditions, concentrating on enriching public space in cities. His focus on the incorporation of art, both in the form of sculptural installations and by the thrust of the landscape design itself, into the public realm is long-standing. He has often described LAB's mission as to produce "art in service to the public." In addition, Mr. Fitch has been active in design education. He has recently taught at the Washington-Alexandria Architecture Consortium, has conducted studios in Harvard University's Graduate School of Design for three years and served on the faculty of the School of Architecture at Howard University for five years.

GRANT ASSOCIATES

Grant Associates is a world-leading British Landscape Architecture consultancy specializing in creative, visionary design of both urban and rural environments worldwide, working with some of the world's leading architects and designers.

Inspired by the connection between people and nature Grant Associates fuses nature and technology in imaginative ways to create cutting edge design built around a concern for the social and environmental quality of life.

Grant Associates has experience in all scales and types of ecological and landscape development including strategic landscape planning, master planning, urban design and regeneration and landscapes for housing, education, sport, recreation, visitor attractions and commerce.

HASSELL

HASSELL is an international design practice with 14 studios in Australia, China, South East Asia and the United Kingdom. With more than 900 people and a track record spanning 75 years, they work globally across a diverse range of markets.

An interdisciplinary practice, they combine expertise in architecture, interior design, landscape architecture and planning with integrated sustainability and urban design capabilities.

As a single, privately owned partnership, each of their studios has the flexibility and autonomy to service both local and global clients, and the advantage of access to their combined resources and collective experience.

INSPIRING PLACE

Inspiring Place creates experiences of place that inspire mind, body and soul - experiences that reveal the essence of a locale whether it be a city, a wilderness reserve or somewhere in between. Founded in 1996 and led by Jerry de Gryse and John Hepper, the Inspiring Place team thrive on delivering insight and creative solutions to questions in the built environment.

Inspiring Place works across the big picture and gets down to the details that matter when it comes to the planning and design of the public realm and the inter-relationships between people and place.

VICENTE GUALLART

Vicente Guallart (b. Valencia 1963) chief architect of the city of Barcelona and general director of Urban Habitat since 2011.

Guallart has been founder of Guallart Architects (1993) and of IAAC (Institute of Advanced Architecture in Catalunya) (2001).

His most relevant and recent projects includes, among others, Sociópolis in Valencia: an innovative housing project for urban and environmental development with projects by international architects; Sharing Bloks in Gandía: a residence for students, the first in Spain where the dynamical relation between private and shared areas can generate a continuous re-configuration and extension of the spaces to live; Fugee Port and Keelung Port in Taiwan. He is author of Geologics (Actar), and co-author of the Metapolis dictionary of Advanced Architecture and Hypercatalonia.

Right now Maria Diaz is in charge of Guallart Architects.

FORM ASSOCIATES

FoRM Associates was established in 2007 by artist Peter Fink, architect Igor Marko and landscape architect Rick Rowbotham to follow their shared concerns in urbanism. The company was formed through an amalgamation of Art2Architecture London Ltd and Urban Red Ltd.

The creation of FoRM Associates has brought together their experiences and skills in urban design, architecture and landscape architecture, to deliver an integrated inter-disciplinary design consultancy for 21st Century cities.

MAYSLITS KASSIF ARCHITECTS

Mayslits Kassif Architects was founded in Tel Aviv by Ganit Mayslits Kassif & Udi Kassif. Since 1994 the practice is involved in a variety of projects in the fields of urban planning, landscape urbanism, public buildings, housing, and retail.

Since 1997 Mayslits Kassif Architects have won several major public competitions such as: Remez-Arlozorov Community Campus in Tel Aviv and the regeneration of Tel Aviv Port public spaces, which won the Rosa Barba European Landscape prize and audience choice in the 6th biennial of Landscape Architecture in Barcelona 2010.

IN SITU

The In Situ agency brings together a team of twelve (landscapers, architects and planners) around Emmanuel Jalbert, landscape and urban planner. In addition to his project management activity, Emmanuel Jalbert works as a consultant and is also involved in teaching and research.

For over 20 years, In Situ has led multiple development projects. It has particular expertise of public space in all its forms: parks, squares, gardens, plazas, docks and riverbanks. Their work extends from dense city to wild landscape and each project aims to combine nature and culture, to reconcile memory and modern times.

SWA GROUP

For over five decades, SWA Group has been recognized as a world design leader in landscape architecture, planning and urban design. Their projects have received over 700 awards and have been showcased in over 60 countries. Their principals are among the industry's most talented and experienced designers and planners. Emerging in 1959 as the West Coast office of Sasaki, Walker and Associates, the firm first assumed the SWA Group name in 1975.

Despite being one of the largest firms of its type in the world, SWA is organized into smaller studio-based offices that enhance creativity and client responsiveness. Over 75% of their work has historically come from repeat clients. In addition to bringing strong aesthetic, functional, and social design ideas to our projects, they're also committed to integrating principles of environmental sustainability. At the core of their work is a passion for imaginative, solution-oriented design that adds value to land, buildings, cities, regions, and to people's lives.

WEISS/MANFREDI ARCHITECTURE/LANDSCAPE/URBANISM

WEISS/MANFREDI is at the forefront of architectural design practices that are redefining the relationships between landscape, architecture, infrastructure, and art. The firm's projects are noted for clarity of vision, bold and iconic forms, and material innovation. Named one of North America's "Emerging Voices" by the Architectural League of New York, WEISS/MANFREDI's distinct vision was recognized in 2004 by the Arts and Letters Award in Architecture from the American Academy of Arts and Letters. Additional honors include the Tau Sigma Delta Gold Medal—an international recognition awarded to one architect annually—and the New York AIA Gold Medal.

ACXT

ACXT was set up as an association of professionals. More than four hundred people work at ACXT today, more than half of them being architects. Engineers are also involved: civil, mechanical, electrical, agricultural..., biologists, experts in acoustics, illumination, telecommunications, costs, building site organization, graphic designers or IT technicians. Due to the problems caused by specialization and the development of technology, they seek solutions that integrate them right from the beginning, in the project conception stage. Knowledge is shared and participation and collaboration is encouraged.

The London magazine Building Design, in its 2012 annual ranking BD World Architecture top 100, placed ACXT in position 48 of the most important firms worldwide by number of architects.

CLAUDE CORMIER + ASSOCIÉS INC

CLAUDE CORMIER + ASSOCIÉS INC. is an internationally recognized practice that extends far beyond the conventional realm of traditional landscape design to forge bridges between urban design, public art, and architecture. The projects of CLAUDE CORMIER + ASSOCIÉS INC. are likewise anything but conventional, celebrating the artificial and surreptitiously altering reality with the surreal. The team work is distinguished not only for its inventiveness – but also its tenacious optimism in the power of design.

Since 1995, the firm has been privileged to work on major public work mostly in Montreal and Toronto.

BURO LUBBERS

Buro Lubbers is a Dutch design studio for landscape architecture, urban planning and the design of public space. Since 1993 the studio has been designing projects at all scale levels, for both private and public commissioners. The studio's projects are characterised by a multidisciplinary approach, innovative expertise, sharp analysis, a sturdy and poetic expression and a critical eye for detail. An enthusiastic, multi-disciplinary team takes care of a total product, from research to preservation. With ambition and a realistic mindset they deploy their creative potential for the building projects of today.

SASAKI ASSOCIATES, INC.

Sasaki was founded 60 years ago on the basis of interdisciplinary planning and design. Today, their services include architecture, interior design, planning, urban design, landscape architecture, strategic planning, civil engineering, and graphic design. Among these disciplines, they collaborate with purpose. Their integrated approach yields rich ideas and surprising insights.

Sasaki is an innovator. They play a leading role in shaping the future of the built environment through bold ideas and new technologies. They approach sustainability through the lenses of economics, social context, and the environment. Their solutions are not only effective but poetic and enduring. Their approach helps clients make smart, long-term decisions that result in greater value for them, and a better future for the planet.

BLACKWELL & ASSOCIATES

Blackwell & Associates was established in 1987. They are a specialised design practice focusing on sustainable landscape architectural and urban design solutions.

Much of their focus since the practice was first established has been on the urban environment and on what people desire and require when they 'trade' a suburban existence for a higher density urban lifestyle.

A strong focus of their practice is in creating environments with a high level of pedestrian amenity, even when they are dealing with streetscapes. In Western Australia, this means placing a strong emphasis on creating attractive, shady streets and public open spaces with high levels of amenity even if such spaces are very simple in form.

Achieving demonstrable, project specific, sustainable outcomes is integral to their design philosophy. In a more holistic context their design objectives revolve around: contributing to community wellbeing; championing environmental stewardship; meeting or exceeding economic expectations and achieving design excellence.

PLACE DESIGN GROUP

PLACE Design Group is an international planning design and environment consultancy with offices in Australia, Asia, the Middle East and Pacific Islands.

Established in 2001, their success has been built on a culture of delivering solutions to planning and design challenges with a combination of innovative and lateral thinking.

Their multidisciplinary approach offers their clients a choice of single or multiple services allowing them to achieve the highest level of project integration.

Their focus on research and development means they regularly create new 'project-specific' initiatives and approaches that ensure the best possible planning and design outcomes are achieved.

AEDAS

Aedas, a leading international design practice, offers services in architecture, interior design, master planning, landscape, urban design and building consultancy within Asia, the Middle East, Europe and the Americas. They are committed to their core value of design excellence and are dedicated to research & development and sustainability. Their 27 global offices allow them to apply international expertise with local knowledge and delivery thus supporting the communities in which they work.

BROOK MCILROY

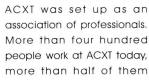

Brook McIlroy is an award-winning architecture, urban design, landscape architecture and planning firm with offices in Toronto and Thunder Bay Canada. Their practice was founded on the ambition to create a truly multi-disciplinary office to address the complex and interrelated challenges of contemporary city building and design.

Projects span from large-scale urban visions and implementation strategies to the detailed design of buildings, landscapes and civic infrastructure. Each project is approached as a unique challenge focused on client engagement, community consultation and thorough place-specific research.

HÄFNER/JIMÉNEZ

On the whole, the landscape architecture of Häfner/Jiménez lays claim to a certain autonomy to the urban and landscaped environment. Important references are seen as obvious, and stressed where needed as an underline of the spatial composition.

Topographic modeling, vegetation backdrops and organizational principals as well as material choice are not only references to 'good' spatial principles. Rather, these elements are used in order to give the future a precise spatial framework. In this way the future is not predestined by the designs of Häfner/Jiménez, their work illustrates an image of the present through the spatial design of the panoramas. Landscape architecture couldn't be more classical.

NYRÉNS ARKITEKTKONTOR

Nyréns Arkitektkontor Nyréns is an award winning Swedish architecture practice where the majority of the employees are shareholders of the company.

The office was founded in 1948 by Carl Nyrén and today they are around 120 people working in two offices in Sweden: in Stockholm and Malmö. Nyréns employs architects, engineers, interior designers, landscape architects, city planners, architectural conservationists, 3D visualisers and architectural modelers. They look for integrated solutions that bring urban design, master planning, building design, engineering, landscaping, interior design and architectural conservationist expertise together.

TAYLOR CULLITY LETHLEAN

T.C.L
TAYLOR.CULLITY.LETHLEAN

Since 1989, Taylor Cullity Lethlean has undertaken an investigation into the poetic expression of the Australian Landscape and contemporary culture. This has permeated their design work in a multiplicity of public settings from urban waterfronts to desert walking trails. In each case the detailed exploration of context, site and community have informed outcomes and enriched the patterning and detail of built landscapes.

The results of this dynamic dialogue with clients, communities, academics, and colleagues is an eclectic body of work woven together by a common thread of quality, commitment, and surprisingly simple but rich environments which support the life of the communities they serve.

WILLIAMS, ASSELIN, ACKAOUI & ASSOCIATES Inc.

WAA is an international firm renowned for its drive and prize-winning projects in landscape architecture, urban planning and urban design.

WAA is known as an innovative leader in the world of design. For over 30 years, WAA's Montreal head office, in Canada, and more recent Shanghai office, in China, have worked on creating a wide variety of quite complex projects.

They develop space intended for citizens and always try to create cultivated environments in which it is pleasant to live. Sustainable development and respect for the environment and local cultures are at the center of their design approach. WAA's international team always work harder to attain excellence in creating urban spaces, public or private.

PWL PARTNERSHIP LANDSCAPE ARCHITECTS INC

PWL Partnership Landscape Architects Inc. is a landscape architectural practice with over thirty-five years of experience in the planning and design of public and private open space. Their depth of expertise and professionalism in British Columbia has afforded them the opportunity to practice their skills in other parts of Canada, the United States, and China.

PWL Partnership's design philosophy draws inspiration from the ecological, historical and cultural aspects of the landscape to produce designs that are innovative, imaginative, sustainable, distinct, appropriate, and cost effective. They attempt to reflect the natural and cultural history of an area in their development of public places. If such references can stir even the slightest curiosity, then a connection is made between the site and the visitor. Such connections turn anonymous sites into memorable places that recognize the local vernacular.

SCAPE LANDSCHAFTSARCHITEKTEN GMBH

scape For the scape office the term "landscape" comprises not only the area of unspoiled nature outside the cities, but all forms of landscape like cityscapes, seascapes and industrial landscapes.

These landscapes are not naturally available, but emerge when a partition of their environment is realized aesthetically. As designers they want to induce, support and direct the processes, which cause development of landscape. The term "scape" stands for designed landscape in the broadest sense: land becomes landscape, city becomes urbanscape, depending on the object, materiality and dimension. If a project shall give impulses for the image of a location, it cannot just satisfy basic requirements. They aim at developing projects for their customers, which offer an outstanding design and allow a diversity of adoptions. Sustainable high-tech solutions result from the use of innovative and ecological methods of construction.

BURO SANT EN CO

BURO
SANT
EN
CO landschapsarchitectuur

Buro Sant en Co was founded in 1990 by Edwin Santhagens and Monique de Vette. The office provides a full range of landscape architecture and urban design services, conceived and executed at the highest artistic level.

Employing a team of skilled designers, each with a different specialization, the company is able to compete and take part in projects at every level, from masterplan to detailed design. The result is a large variety of realized projects such as: parks, campuses, squares, shopping areas, boulevards, gardens and estates.

Designing is the search for distinctive and sensitive solutions to individual problems. The style of design is important. They aim to make innovative plans that have a certain beauty and artistic expression. Yet functionality and maintenance are just as important. They spent great effort on detailing and material choice. Functionality, simplicity and sustainability are keywords for their design approach.

MICHEL CORAJOUD

Consider as one of the founders of the revival of the landscape career, Michel Corajoud asserted himself by a strong personality. It is in the struggle against the trend of his elder landscape architects to want "to neutralize" the city, that he based at the same time his thought and his practice.

His practice grew rich of new notions: that of the interrelation. It is to pursue the activities and the philosophie developed that he created in 2008 the CORAJOUD-SALLIOT-TABORDA workshop, on association with Yannick Salliot and José Luis Taborda. The CORAJOUD-SALLIOT-TABORDA workshop, conceives and realizes studies of town planning, studies of landscape, urban development.

EAA-EMRE AROLAT ARCHITECTS

Emre Arolat Architects was founded in May 2004 by Emre Arolat and Gonca Paþolar in Istanbul, as the continuation of Emre Arolat's architectural practices which he started at Arolat Architects as an associate designer in 1987. Arolat Architects was founded in 1961 and the group has drawn significant professional experience from regular contributions to architectural competitions and executed significant housing, tourism facilities, leisure centers, administrative buildings and sports grounds projects. The practice is being continued with the same range of projects in EAA-Emre Arolat Architects with the contributions of other partners, Neþet Arolat, Þaziment Arolat, Sezer Bahtiyar and a professional architectural staff in its office in Istanbul.

ACKNOWLEDGEMENTS

We would like to thank everyone involved in the production of this book, especially all the artists, designers, architects and photographers for their kind permission to publish their works. We are also very grateful to many other people whose names do not appear on the credits but who provided assistance and support. We highly appreciate the contribution of images, ideas, and concepts and thank them for allowing their creativity to be shared with readers around the world.